BIOLOGY ANTHOLOGY

READINGS IN THE LIFE SCIENCES

SAMUEL WILSON and RICHARD L. ROE

WEST PUBLISHING CO.
St. Paul • New York • Boston
Los Angeles • San Francisco

COPYRIGHT © 1974 By WEST PUBLISHING CO.
All rights reserved
Printed in the United States of America

Library of Congress Catalog Number: 74–2810
ISBN: 0–8299–0019–5
  Wilson & Roe—Biology Anthology

PREFACE

There are many avenues to involvement with the life sciences.

Albert Szent-Gyorgyi is a rebel. He is an accomplished analyst of life, but his ideas often dispute mainstreams of biological thought.

John Steinbeck was a romantic. He looked for beauty and truth in the tide pools of the Sea of Cortez.

Jan Van Lawich-Goodall empathizes with her fellow beings.

Pierre Teilhard de Chardin was both a scientist and a mystic in his vision of the future of man.

George Bernard Shaw gave us humor with his revelation of the "Life Force". He saw evolution ultimately leading to existence for the contemplation of life.

We hope that this anthology will help you in your search for the truths of existence. But, we stress that you shouldn't believe anything that you read. Rather, look for sense (and nonsense).

Sam Wilson

Richard L. Roe

*

ACKNOWLEDGEMENTS

Many people helped us in compiling this anthology. Professor Bill Wilson of San Diego State University analyzed the initial proposal, and helped give direction to our literature search. Gary Swartz gave us invaluable aesthetic advice. Professors J. Martin Ikkanda of Los Angeles Pierce College, Phil Powers and Don Villeneuve of Ventura College and Charles Dreiling of the University of Nevada, Reno reviewed potential selections, and helped us in choosing from the numerous possibilities. Thanks also to our friends who told us what they found interesting in the study of life.

A final note of appreciation to Clyde Perlee Jr., Editor, and Kenneth Zeigler, Publisher, for their encouragement.

*

TABLE OF CONTENTS

PART I - BIOLOGY IN PERSPECTIVE

    "Psychology", Aristotle 3

    "Of the Schematisme or Texture of Cork, and of the Cells and Pores of Some Other Such Frothy Bodies", Robert Hooke . 5

    "An Essay on the Principle of Population", Robert Malthus 7

    "Letter to Asa Gray, September 5, 1875", Charles Darwin . . 13

    A Foreword to the Popular Edition of "Man and Superman", George Bernard Shaw 17

    "Bryan and Darrow at Dayton", Leslie H. Allen 23

    "The Log from the Sea of Cortez", John Steinbeck 29

    "Another Bad Trip", Science News 37

    "Brave New World Revisited", Aldous Huxley 39

    "The Tragedy of the Commons", Garrett Hardin 45

    "Soured on Sugar", Daniel Henninger 55

PART II - EVOLUTION

    "Eden and Evolution", Ron Moskowitz 59

    "Land Life: A Speculation", Isaac Asimov 63

    "African Fossil Bonanza Points to Earliest Man", Stanley Meisler 65

    "Drugs May be Unable to Control Some Killer Diseases, Experts Say" 67

    "What Aggression is Good For", Konrad Lorenz 69

    Two Poems, William Blake 83

    "The End of the Species", Pierre Teilhard de Chardin . . . 85

PART III - ECOLOGY

"The Historical Roots of Our Ecologic Crisis", Lynn White, Jr. 91

"Future is Famine for Ten Million South of Sahara" 99

"A-Bombs, Bugbombs, and Us", G. M. Woodwell, W. M. Malcolm, R. H. Whittaker 105

"Shippingport - The Killer Reactor?", Richard Lewis 113

"Sunlight and the SST" 117

"Another Pollution Danger: Too Few Ions" 119

"Mystery Disease Arrives Quietly, Stuns Workers, Don Frederick, Jr. 121

"Power Plants and Cottontails", Bruce Wallace 127

"Mosquitoes Face Deadly Foe" 131

"Infectious Cure", Kevin P. Shea 133

"Aussies Turn to Beetles in Ecology Fight", David Lamb . . 137

"The Case of the Chinese Clams: What to do?" 141

"Man and Nature: Symbiosis" 143

PART IV - MOLECULAR BIOLOGY

"Vegetarianism: Fad, Faith, or Fact?", Sanat K. Majumder 147

"Diet for a Small Planet", Frances Moore Lappe 155

"Peanut Butter Protein" 167

"Stricter Controls on Vitamins" 169

"Milk and Blindness in Brazil", G. Edwin Bunce 173

"Vitamins and the Fetus: The benefits of B_{12}". 177

"Hot Dogs and Hyperkinesis" 179

"Food labels to say more about nutrition" 181

"Birth Control: Current Technology, Future Prospects", Jean L. Marx 183

PART IV - MOLECULAR BIOLOGY

"The Panacea Pill" 189

"Breakfast of Champions", Kurt Vonnegut, Jr. 191

"Will vitamins replace the psychiatrist's couch?",
Robert J. Trotter 193

"Knots", R. D. Laing 197

PART V - GENETICS

"Creeping Up on Eugenics", Phillip Chapnick 201

"Precautions for Prospective Parents" 207

"Doomed to Inequality?", Gerald Chasin 210

"Board of Education Rejects Concept of Racial
Superiority" 215

"A 126-unit artificial gene" 217

"Molecular Rhythms", Payson Stevens 220

"Introduction to Biology Today", Albert Szent-Gyorgyi . . . 221

PART VI - STRUCTURE AND FUNCTION

"The Longevity Seekers", Albert Rosenfeld 225

"Shuffle Brain", Paul Pietsch 233

"No Blinkers for the Brain", Professor John Taylor 243

"A Heart-Stopping, Eye-Bulging, Wave-Making Idea",
Scot Morris 247

"Gadget Keeps the Tummy Tight" 259

"Out on a Lymph with Burt Bacteria" 261

"Stalking the Wild Crown Gall", Wallace Cloud 269

"Cultured Organs" 277

"Triple Transplant Donor Slaying Dilemma" 279

"Medical ethics and human subjects" 281

PART VI - STRUCTURE AND FUNCTION

    "Tests on Aborted Live Fetuses Banned" 285

    "Acupuncture: The World's Oldest System of Medicine",
    John W. White 287

PART VII - BEHAVIOR

    "Flo's Sex Life", Jane Goodall 297

    "The biological depths of loneliness", Robert J.
    Trotter 303

    "Conscription at Sea", Bruce Wallace 307

    "Tuning in on Porpoises that "work" and "talk" with
    people 311

    "The scarred species", Patricia McBroom 317

PART VIII - NATURAL HISTORY

    "The Bison's Woe", Kevin P. Shea 323

    "Economics of Wild Strawberries", Euell Gibbons 329

    "Dandelion Greens I". 335

    "Paddlefish Cultivation Possible", Col. George S. Bumpas . 337

PHOTOGRAPH CREDITS

Eschen Foundation--Haags Gemeentmuseum, The Hague
    Pages 1, 59, 89, 145, 199, 223, 295, 321

United Press International
    Pages 17, 25, 29, 259, 323

U. S. Department of Agriculture
    Pages 134, 135

Official U. S. Navy Photographs
    Pages 311, 312, 313

THE BIOLOGY ANTHOLOGY

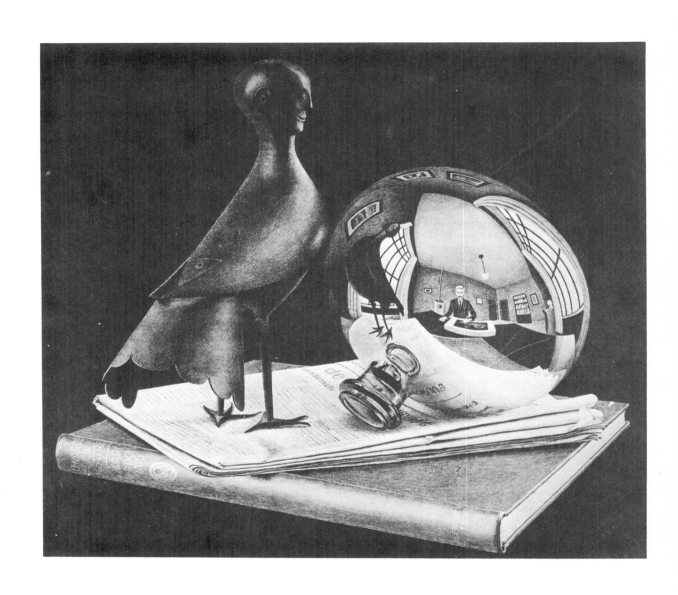

PART I
BIOLOGY IN PERSPECTIVE

PSYCHOLOGY

Aristotle

Aristotle, who founded the Lyceum in Athens in 334 BC, was among the first who's thoughts on the life sciences were written. During the last ten years of his life, he was constantly surrounded by students at the Lyceum, who recorded nearly his every word. Although many of his conclusions have been proved incorrect, his ideas are a precious record of the state of development of human awareness in the mystical culture of early Greece. We wonder if (in spite of the incredible expansion of knowledge since the dawn of history) mankind has lost truths known to his ancient ancestors.

The following includes some of Aristotle's thoughts on sensory stimulation.

Sensation, ..., consists in being moved and acted upon; and it is commonly regarded as a sort of qualitative change. There are some who add that all change is an action of like upon like. To what extent this principle can be accepted we have explained in our general discussion of activity and passivity.

It may be wondered why sensation does not arise in the sense-organs, themselves. These are known to contain fire and earth and the other elements, which either in themselves or through their attributes are the objects of sense-experience: why, then can they not produce sensations without the presence of external objects ? The answer must be that sensation is present in them not actually but only potentially: in the same way that combustible objects obviously cannot catch fire of themselves without anything to ignite them, for if they could they would not need the application of already existing fire to set them ablaze. Sensation thus has two meanings: we attribute seeing and hearing to what has the capacity to see and hear, though perhaps momentarily asleep, no less than to what is actually seeing and hearing at the moment. This indicates the twofold meaning of sensation, as potential and actual; and 'to be sentient' likewise has both these meanings.

OF THE SCHEMATISME OR TEXTURE OF CORK, AND OF THE CELLS AND PORES OF SOME OTHER SUCH FROTHY BODIES

Robert Hooke

Robert Hooke's use of the microscope was an early landmark in the development of the technology of biological observation. Throughout history, mankind's accellerating sophistication in tool creation has fanned the fires of scientific capability. Scientific thought, in turn, has provided the roots of our technology.

I took a good clear piece of Cork, and with a Pen-knife sharpened as keen as a Razor, I cut a piece of it off, and thereby left the surface of it exceeding smooth, then examining it very diligently with a Microscope, me thought I could perceive it to appear a little porous; but I could not so plainly distinguish them, as to be sure that they were pores, much less what Figure they were of: But judging from the lightness and yielding quality of the Cork, that certainly the texture could not be so curious, but that possibly, if I could use some further diligence, I might find it to be discernible with a Microscope, I with the same sharp Pen-knife, cut off from the former smooth surface an exceeding thin piece of it, and placing it on a black object Plate, because it was itself a white body, and casting the light on it with a deep plano-convex Glass, I could exceeding plainly perceive it to be all perforated and porous, much like a Honey-comb, but that the pores of it were not regular; yet it was not unlike a Honey-comb in these particulars.

First, in that it had a very little solid substance, in comparison of the empty cavity that was contained between, as does more manifestly appear by the

---

From *Micrographia*, Martin and Allestry, London, 1665.

Figure A and B of the XI Scheme, for the Interstitia, or walls (as I may so call them) or partitions of those pores were near as thin in proportion to their pores, as those thin films of Wax in a Honey-comb (which enclose and constitute the sexangular cells) are to theirs.

Next, in that these pores, or cells, were not very deep, but consisted of a great many little Boxes, separated out of one continued long pore, by certain Diaphragms, as is visible by the Figure B, which represents a sight of those pores split the long-ways.

I no sooner discern'd these (which were indeed the first microscopical pores I ever saw, and perhaps, that were ever seen, for I had not met with any Writer or Person, that had made any mention of them before this) but me thought I had with the discovery of them, presently hinted to me the true and intelligible reason of all the Phaenomena of Cork; As,

First, if I enquir'd why it was so exceeding light a body? my Microscope could presently inform me that here was the same reason evident that there is found for the lightness of froth, an empty Honey-comb, Wool, a Spunge, a Pumice-stone, or the like; namely, a very small quantity of a solid body, expended into exceeding large dimensions.

AN ESSAY ON THE PRINCIPLE OF POPULATION
Robert Malthus

    The dilema of over-population, that Thomas Robert Malthus revealed in 1798, has spawned many of today's environmental problems. In simple terms, if the people of the technologically advanced nation don't hold back on material consumption, and if the people of the less developed nations don't slow down their reproduction, then warfare over limited resources will certainly continue, with increasing fury.

    Apart from the intrinsic value of Malthus's essay, it was credited by Charles Darwin as having helped him to perceive the principle of natural selection. In Darwin's own words, "I happened to read for amusement Malthus on Population, and being well prepared to appreciate the struggle for existence which everywhere goes on, from long-continued observation in the habits of animals and plants, it at once struck me that under these circumstances favorable variations would tend to be preserved and unfavorable ones to be destroyed".

    In an inquiry concerning the improvement of society, the mode of conducting the subject which naturally presents itself, is,

1. To investigate the causes that have hitherto impeded the progress of mankind towards happiness; and,
2. To examine the probability of the total or partial removal of these causes in future.

From *An Essay on the Principle of Population*, London, 1798.

To enter fully into this question, and to enumerate all the causes that have hitherto influenced human improvement, would be much beyond the power of an individual. The principal object of the present essay is to examine the effects of one great cause intimately united with the very nature of man; which, though it has been constantly and powerfully operating since the commencement of society, has been little noticed by the writers who have treated this subject. The facts which establish the existence of this cause have, indeed, been repeatedly stated and acknowledged; but its natural and necessary effects have been almost totally overlooked; though probably among these effects may be reckoned a very considerable portion of that vice and misery, and of that unequal distribution of the bounties of nature, which it has been the unceasing object of the enlightened philanthropist in all ages to correct.

The cause to which I allude is the constant tendency in all animated life to increase beyond the nourishment prepared for it.

It is observed by Dr. Franklin that there is no bound to the prolific nature of plants or animals but what is made by their crowding and interfering with each other's means of subsistence. Were the face of the earth, he says, vacant of other plants, it might be gradually sowed and overspread with one kind only, as for instance with fennel: and were it empty of other inhabitants, it might in a few ages be replenished from one nation only, as for instance with Englishmen.

This is incontrovertibly true. Through the animal and vegetable kingdoms Nature has scattered the seeds of life abroad with the most profuse and liberal hand; but has been comparatively sparing in the room and the nourishment necessary to rear them. The germs of existence contained in this earth, if they could freely develop themselves, would fill millions of worlds in the course of a few thousand years. Necessity, that imperious, all pervading law of nature, restrains them within the prescribed bounds. The race of plants and the race of animals shrink under this great restrictive law; and man cannot by any efforts of reason escape from it.

In plants and irrational animals, the view of the subject is simple. They are all impelled by a powerful instinct to the increase of their species; and this instinct is interrupted by no doubts about providing for their offspring. Wherever therefore there is liberty, the power of increase is exerted; and the superabundant effects are repressed afterwards by want of room and nourishment.

The effects of this check on man are more complicated. Impelled to the increase of his species by an equally powerful instinct, reason interrupts his areer, and asks him whether he may not bring beings into the world for whom he cannot provide the means of support. If he attend to this natural suggestion, the restriction too frequently produces vice. If he hear it not, the human race will be constantly endeavouring to increase beyond the means of subsistence. But as, by the law of our nature which makes food necessary to the life of man, population can never actually increase beyond the lowest nourishment capable of supporting it, a strong check on population, from the difficulty of acquiring food, must be constantly in operation. This difficulty must fall somewhere, and must necessarily be severely felt in some or other of the various forms of misery, or the fear of misery, by a large portion of mankind.

That population has this constant tendency to increase beyond the means of subsistence, and that it is kept to its necessary level by these causes, will sufficiently appear from a review of the different states of society in which man has existed. But, before we proceed to this review, the subject will, perhaps, be seen in a clearer light if we endeavour to ascertain what would be the natural increase of population if left to exert itself with perfect freedom; and what might be expected to be the rate of increase in the productions of the earth under the most favourable circumstances of human industry.

It will be allowed that no country has hitherto been known where the manners were so pure and simple, and the means of subsistence so abundant, that no check whatever has existed to early marriages from the difficulty of providing for a family, and that no waste of the human species has been occasioned by vicious customs, by towns, by unhealthy occupations, or too severe labour. Consequently in no state that we have yet known has the power of population been left to exert itself with perfect freedom.

Whether the law of marriage be instituted, or not, the dictate of nature and virtue seems to be an early attachment to one woman; and where there were no impediments of any kind in the way of an union to which such an attachment would lead, and no causes of depopulation afterwards, the increase of the human species would be evidently much greater than any increase which has been hitherto known.

In the northern states of America, where the means of subsistence have been more ample, the manners of the people more pure, and the checks to early marriages fewer than in any of the modern states of Europe, the population has been found to double itself, for above a century and a half successively, in less than twenty-five years. Yet, even during these periods, in some of the towns, the deaths exceeded the births, a circumstance which clearly proves that, in those parts of the country which supplied this deficiency, the increase must have been much more rapid than the general average.

In the back settlements, where the sole employment is agriculture, and vicious customs and unwholesome occupations are little known, the population has been found to double itself in fifteen years. Even this extraordinary rate of increase is probably short of the utmost power of population. Very severe labour is requisite to clear a fresh country; such situations are not in general considered as particularly healthy; and the inhabitants, probably, are occasionally subject to the incursions of the Indians, which may destroy some lives, or at any rate diminish the fruits of industry.

According to a table of Euler, calculated on a mortality of 1 in 36, if the births be to the deaths in the proportion of 3 to 1, the period of doubling will be only 12 years and 4-5ths. And this proportion is not only a possible supposition, but has actually occurred for short periods in more countries than one.

Sir William Petty supposes a doubling possible in so short a time as ten years.

But, to be perfectly sure that we are far within the truth, we will take the slowest of these rates of increase, a rate in which all concurring testimonies agree, and which has been repeatedly ascertained to be from procreation only.

It may safely be pronounced, therefore, that population, when unchecked, goes on doubling itself every twenty-five years, or increases in a geometrical ratio.

The rate according to which the productions of the earth may be supposed to increase, it will not be so easy to determine. Of this, however, we may be perfectly certain, that the ratio of their increase in a limited territory must be of a totally different nature from the ratio of the increase of population. A thousand millions are just as easily doubled every twenty-five years by the power of population as a thousand. But the food to support the increase from the greater number will by no means be obtained with the same facility. Man is necessarily confined in room. When acre has been added to acre till all the fertile land is occupied, the yearly increase of food must depend upon the melioration of the land already in possession. This is a fund, which, from the nature of all soils, instead of increasing, must be gradually diminishing. But population, could it be supplied with food, would go on with unexhausted vigour; and the increase of one period would furnish the power of a greater increase the next, and this without any limit.

From the accounts we have of China and Japan, it may be fairly doubted whether the best - directed efforts of human industry could double the produce of these countries even once in any number of years. There are many parts of the globe, indeed, hitherto uncultivated, and almost unoccupied; but the right of exterminating, or driving into a corner where they must starve, even the inhabitants of these thinly-peopled regions, will be questioned in a moral view. The process of improving their minds and directing their industry would necessarily be slow; and during this time, as population would regularly keep pace with the increasing produce, it would rarely happen that a great degree of knowledge and industry would have to operate at once upon rich unappropriated soil. Even where this might take place, as it does sometimes in new colonies, a geometrical ratio increases with such extraordinary rapidity, that the advantage could not last long. If the United States of America continue increasing, which they certainly will do, though not with the same rapidity as formerly, the Indians will be driven further and further back into the country, till the whole race is ultimately exterminated, and the territory is incapable of further extension.

These observations are, in a degree, applicable to all the parts of the earth where the soil is imperfectly cultivated. To exterminate the inhabitants of the greatest part of Asia and Africa is a thought that could not be admitted for a moment. To civilise and direct the industry of the various tribes of Tartars and Negroes would certainly be a work of considerable time and of variable and uncertain success.

Europe is by no means so fully peopled as it might be. In Europe there is the fairest chance that human industry may receive its best direction. The science of agriculture has been much studied in England and Scotland; and there is still a great portion of uncultivated land in these countries. Let us consider at what rate the produce of this island might be supposed to increase under circumstances the most favourable to improvement.

If it be allowed that by the best possible policy, and great encouragements to agriculture, the average produce of the island could be doubled in the first twenty-five years, it will be allowing, probably, a greater increase than could with reason be expected.

In the next twenty-five years, it is impossible to suppose that the produce could be quadrupled. It would be contrary to all our knowledge of the properties of land. The improvement of the barren parts would be a work of time and labour; and it must be evident to those who have the slightest acquaintance with agricultural subjects that, in proportion as cultivation extended, the additions that could yearly be made to the former average produce must be gradually and regularly diminishing. That we may be the better able to compare the increase of population and food, let us make a supposition, which, without pretending to accuracy, is clearly more favourable to the power of production in the earth than any experience we have had of its qualities will warrant.

Let us suppose that the yearly additions which might be made to the former average produce, instead of decreasing, which they certainly would do, were to remain the same; and that the produce of this island might be increased every twenty-five years by a quantity equal to what it at present produces. The most enthusiastic speculator cannot suppose a greater increase than this. In a few centuries it would make every acre of land in the island like a garden.

If this supposition be applied to the whole earth, and if it be allowed that the subsistence for man which the earth affords might be increased every twenty-five years by a quantity equal to what it at present produces, this will be supposing a rate of increase much greater than we can imagine that any possible exertions of mankind could make it.

It may be fairly pronounced, therefore, that, considering the present average state of the earth, the means of subsistence, under circumstances the most favourable to human industry, could not possibly be made to increase faster than in an arithmetical ratio.

The necessary effects of these two different rates of increase, when brought together, will be very striking. Let us call the population of this island eleven million; and suppose the present produce equal to the easy support of such a number. In the first twenty-five years the population would be twenty-two millions, and the food being also doubled, the means of subsistence would be equal to this increase. In the next twenty-five years, the population would be forty-four millions, and the means of subsistence only equal to the support of thirty-three millions. In the next period the population would be eighty-eight millions, and the means of subsistence just equal to the support of half that number. And, at the conclusion of the first century, the population would be a hundred and seventy-six millions, and the means of subsistence only equal to the support of fifty-five millions, leaving a population of a hundred and twenty-one millions totally unprovided for.

Taking the whole earth, instead of this island, emigration would of course be excluded; and, supposing the present population equal to a thousand millions, the human species would increase as the numbers, 1, 2, 4, 8, 16, 32, 64, 128, 256, and subsistence as 1, 2, 3, 4, 5, 6, 7, 8, 9. In two centuries the population would be to the means of subsistence as 256 to 9; in three centuries as 4096 to 13, and in two thousand years the difference would be almost incalculable.

In this supposition no limits whatever are placed to the produce of the earth. It may increase for ever and be greater than any assignable quantity; yet still the power of population being in every period so much superior, the increase of the human species can only be kept down to the level of the means of subsistence by the constant operation of the strong law of necessity, acting as a check upon the greater power.

*

LETTER TO ASA GRAY, SEPTEMBER 5, 1875

Charles Darwin

Charles Darwin had a ponderous intellect. His book, <u>The Origin of Species</u>, is an enormous compendium supporting the process of evolution. His tedious approach was necessary, however, to sway intellectuals to the support of this religiously controversial theory.

The concept of evolution had been around for a long time when Darwin published his book. His creativity was expressed, not as a revelation of evolution, but in his perception of the cutting edge of the evolutionary force--natural selection. His ideas about natural selection are expressed in this letter to a friend.

The post script to the letter summarizes Darwin's impass on the question of variability. Ironically, a contemporary of Darwin, Gregor Mendel, was making unpublicized discoveries on this very problem. Mendel's experiments with pea plants were the unheralded beginnings of the science of genetics.

Down, Sept. 5th

My dear Gray,--I forget the exact words which I used in my former letter, but I dare say I said that I thought you would utterly despise me when I told you what views I had arrived at, which I did because I thought I was bound as an honest man to do so. I should have been a strange mortal, seeing how much I owe to your quite extraordinary kindness, if in saying this I had meant to attribute the least bad feeling to you. Permit me to tell you that, before I had ever corresponded with

---

Reprinted from *The Life and Letters of Charles Darwin*, edited by Francis Darwin, London, 1888.

you, Hooker had shown me several of your letters (not of a private nature), and these gave me the warmest feeling of respect to you; and I should indeed be ungrateful if your letters to me, and all I have heard of you, had not strongly enhanced this feeling. But I did not feel in the least sure that when you knew whither I was tending, that you might not think me so wild and foolish in my views (God knows, arrived at slowly enough, and I hope conscientiously), that you would think me worth no more notice or assistance. To give one example: the last time I saw my dear old friend Falconer, he attacked me most vigorously, but quite kindly, and told me, "You will do more harm than any ten Naturalists will do good. I can see that you have already <u>corrupted</u> and half-spoiled Hooker!!" Now when I see such strong feeling in my oldest friends, you need not wonder that I always expect my views to be received with contempt. But enough and too much of this.

I thank you most truly for the kind spirit of your last letter. I agree to every word in it, and think I go as far as almost anyone in seeing the grave difficulties against my doctrine. With respect to the extent to which I go, all the arguments in favour of my notions fall <u>rapidly</u> away, the greater the scope of forms considered. But in animals, embryology leads me to an enormous and frightful range. The facts which kept me longest scientifically orthodox are those of adaptation--the pollen-masses in asclepias--the mistletoe, with its pollen carried by insects, and seed by birds--the woodpecker, with its feet and tail, beak and tongue, to climb the tree and secure insects. To talk of climate or Lamarckian habit producing such adaptations to other organic beings is futile. This difficulty I believe I have surmounted. As you seem interested in the subject, and as it is an <u>immense</u> advantage to me to write to you and to hear, even so briefly, what you <u>think</u>, I will enclose (copied, so as to save you trouble in reading) the briefest abstract of my notions on the means by which Nature makes her species. Why I think that species have really changed, depends on general facts in the affinities, embryology, rudimentary organs, geological history, and geographical distribution of organic beings. In regard to my Abstract, you must take immensely on trust, each paragraph occupying one or two chapters in my book. You will, perhaps, think it paltry in me, when I ask you not to mention my doctrine; the reason is, if any one, like the author of the 'Vestiges,' were to hear of them, he might easily work them in, and then I should have to quote from a work perhaps despised by naturalists, and this would greatly injure any chance of my views being received by those alone whose opinions I value...

I. It is wonderful what the principle of Selection by Man, that is the picking out of individuals with any desired quality, and breeding from them, and again picking out, can do. Even breeders have been astonished at their own results. They can act on differences inappreciable to an uneducated eye. Selection has been <u>methodically</u> followed in Europe for only the last half century. But it has occasionally, and even in some degree methodically, been followed in the most ancient times. There must have been also a kind of unconscious selection from the most ancient times, namely, in the preservation of the individuals animals (without any thought of their offspring) most useful to each race of man in his particular circumstances. The "roguing," as nursery-men call the destroying of varieties, which depart from their type, is a kind of selection. I am convinced that intentional and occasional selection has been the main agent in making our domestic races. But, however this may be, its great power of modification has been indisputedly shown in late times. Selection acts only by the accumulation of every slight or greater variations, caused by external conditions, or by the mere fact that in generation the child is absolutely similar to its parent. Man, by his power of accumulating variations, adapts living beings to his wants--he may be said to make the wool of one sheep good for carpets, and another for cloth, &c.

II. Now, suppose there was a being, who did not judge by mere external appearance, but could study the whole internal organisation—who never was capricious—who should go on selecting for one end during millions of generations, who will say what he might not effect! In nature we have some <u>slight</u> variations, occasionally in all parts: and I think it can be shown that a change in the conditions of existence is the main cause of the child not exactly resembling its parents; and in nature, geology shows us what changes have taken place, and are taking place. We have almost unlimited time: no one but a practical geologist can fully appreciate this: think of the Glacial period, during the whole of which the same species of shells at least have existed; there must have been, during this period, millions on millions of generations.

III. I think it can be shown that there is such an unerring power at work, or <u>Natural Selection</u> (the title of my book), which selects exclusively for the good of each organic being. The elder De Candolle, W. Herbert, and Lyell, have written strongly on the struggle of life; but even they have not written strongly enough. Reflect that every being (even the elephant) breeds at such a rate that, in a few years, or at most a few centuries or thousands of years, the surface of the earth would not hold the progeny of any one species. I have found it hard constantly to bear in mind that the increase of every single species is checked during some part of its life, or during some shortly recurrent generation. Only a few of those annually born can live to propagate their kind. What a trifling difference must often determine which shall survive and which perish!

IV. Now take the case of a country undergoing some change; this will tend to cause some of its inhabitants to vary slightly; not but what I believe most beings vary at all times enough for selection to act on. Some of its inhabitants will be exterminated and the remainder will be exposed to the mutual action of a different set of inhabitants, which I believe to be more important to the life of each being than mere climate. Considering the infinitely various ways beings have to obtain food by struggling with other beings, to escape danger at various times of life, to have their eggs or seeds disseminated, &c., &c., I cannot doubt that during millions of generations individuals of a species will be born with some slight variation profitable to some part of its economy; such will have a better chance of surviving, propagating this variation, which again will be slowly increased by the accumulative action of natural selection; and the variety thus formed will either coexist with, or more commonly will exterminate its parent form. An organic being like the woodpecker, or the mistletoe, may thus come to be adapted to a score of contingencies; natural selection, accumulating those slight variations in all parts of its structure which are in any way useful to it, during any part of its life.

V. Multiform difficulties will occur to every one on this theory. Most can, I think, be satisfactorily answered.—"Natura non facit saltum" answers some of the most obvious. The slowness of the change, and only a very few undergoing change at any one time answers others. The extreme imperfections of our geological records answers others.

VI. One other principle, which may be called the principle of divergence, plays, I believe, an important part in the origin of species. The same spot will support more life if occupied by very diverse forms: we see this in the many generic forms in a square yard of turf (I have counted twenty species belonging to eighteen genera), or in the plants and insects, on any little uniform islet, belonging to almost as many genera and families as to species. We can understand this with the higher animals, whose habits we best understand. We know that it has been experimentally shown that a plot of land will yield a greater weight, if cropped with several species of grasses, than with two or three species. Now every single organic being, by propagating rapidly, may be said to be striving its utmost to increase numbers. So it will be with the offspring of any species after it has broken into varieties, or sub-species, or true species. And it follows, I think, from the foregoing facts, that the varying offspring of each species will try (only a few will succeed) to seize on as many and as diverse places in the economy of nature as possible. Each new variety or species when formed will generally take the place of, and so exterminate, its less well-fitted parent. This, I believe, to be the origin of the classification or arrangement of all organic beings at all times. These always <u>seem</u> to branch and sub-branch like a tree from a common trunk; the flourishing twigs, destroying the less vigrous--the dead and lost branches rudely representing extinct genera and families.

This sketch is <u>most</u> imperfect; but in so short a space I cannot make it better. Your imagination must fill up many wide blanks. Without some reflection, it will appear all rubbish; perhaps it will appear so after reflection.

C. D.

P.S.--This little abstract touches only the accumulative power of natural selection, which I look at as by far the most important element in the production of new forms. The laws governing the incipient of primordial variation (unimportant except as the groundwork for selection to act on, in which respect it is all important), I shall discuss under several heads, but I can come, as you may well believe, only to very partial and imperfect conclusions.

1857

Reprinted from <u>The Life and Letters of Charles Darwin</u>, ed. by Francis Darwin, London, 1888.

George Bernard Shaw

A Foreword to the Popular Edition

MAN AND SUPERMAN
George Bernard Shaw

George Bernard Shaw was among the first and most eloquent of the creative writers who have let their minds play with the implications of evolution. In the following forword to "Man and Superman" he states the purpose of his most bold vision of the future of man.

I have a special reason for making this book more widely accessible than it can be, ..., at the customary price of my books. It is described on the title page as "A Comedy and a Philosophy." It might have been called a religion as well; for the vision of hell in the third act, which has never been performed on the stage except as a separate work, is expressly intended to be a revelation of the modern religion of evolution.

Putting it roughly and briefly, the discovery and vogue of evolution, from 1790 to 1830, made an end, for the pioneers of thought, of XVIII century rationalist atheism and deism on the one hand, and, on the other, of what may be called Garden-of-Edenism. In the middle of the XIX century the discovery by Darwin and Wallace of Natural Selection, which might have had as its sub-title "The revelation of a method by which all the appearances of intelligent design in the universe may have been produced by pure accident," practically destroyed religion in cultured Europe for a whole generation, in spite of Darwin's and Wallace's own reservations and the urgent warnings and fierce criticisms of Samuel Butler. With the beginning of the present century the return swing of the pendulum has inaugurated a counterfashion of saying that Natural Selection, instead of accounting for everything, accounts

---

Reprinted by permission of The Society of Authors, on behalf of the Bernard Shaw Estate.

for nothing. That it accounts for nothing in any religious sense is of course true; for it leaves untouched the whole sphere of will, purpose, design, intention, even consciousness; and a religion is nothing but a common view of the nature of will, the purpose of life, the design of organism, and the intention of evolution. Such a common view has been gradually detaching itself from the welter of negation provoked by the extremely debased forms of religion which have masqueraded as Christianity in England during the period of petty commercialism from which we are emerging. The time has come for an attempt to formulate this common view as a modern religion, and to provide it with a body of doctrine, a poesy, and a political and industrial system. Shelley and Wagner made notable attempts to provide it with materials for a Bible; and I, with later lights of science to guide me than either of these prophets had, have made a further attempt in this Man and Superman. As I have not been sparing of such lighter qualities as I could endow the book with for the sake of those who ask nothing from a play but an agreeable pastime, I think it well to affirm plainly that the third act, however fantastic its legendary framework may appear, is a careful attempt to write a new Book of Genesis for the Bible of the Evolutionists; and as it is recognized that no matter how highly priced other works of art may be, Bibles must be cheap, I have consented to the issue of this edition at the lowest price possible in the book market to-day.

22nd March, 1911   G.B.S.

    To set out to write a new Book of Genesis is no small undertaking. But, George Bernard Shaw showed little doubt in his abilities. Shaw's humor and insight give a beautiful dimension to the contemplation of the theory of evolution.
    In "Man and Superman", Shaw dramatized his vision of the "life-force" -- an evolutionary force driving life to ever greater powers of self-contemplation. Although the medium of the comedy is distant from the logical, yet poetic, pronouncements of Pierre Teilhand de Chardin (see "The End of the Species"), it appears to us that both writers had a similar vision of the future of man. The following excerpts from Shaw's play are uniquely fascinating, in that his perception of the male and female evolutionary roles seem relevant to the feminist movement.
    Because limitations in space prevent our reproducing the whole third act, a short description of the setting and characters may help in your enjoyment of the excerpts. The scene is Hell. Four characters take part in the exchanges:

> DON JUAN. This character, besides representing the infamous Spanish lover of the 16th century, is also the personification (in death) of a living character, John Tanner, who appears elsewhere in the play. It is probably safe to assume that Don Juan's words reflect Shaw's ideas.
> ANA. She has just entered Hell, and has been revealed to be the personification of John Tanner's lover, Ann Whitefield.
> THE STATUE. He resides in Heaven, but has come to Hell on one of his periodic visits. In life, he was Ann

Whitefield's father, who was killed in a duel with John Tanner.
THE DEVIL. This character also has a living counterpart, Mendoza, who appears elsewhere in the ethereal third act, but not in the rest of the play.

DON JUAN. I enjoy the contemplation of that which interests me above all things: namely, Life: the force that ever strives to attain greater power of contemplating itself. What made this brain of mine, do you think? Not the need to move my limbs; for a rat with half my brains moves as well as I. Not merely the need to do, but the need to know what I do, lest in my blind efforts to live I should be slaying myself.

THE DEVIL. And is Man any the less destroying himself for all this boasted brain of his? Have you walked up and down upon the earth lately? I have; and I have examined Man's wonderful inventions. And I tell you that in the arts of death he outdoes Nature herself, and produces by chemistry and machinery all the slaughter of plague, pestilence, and famine. The peasant I tempt today eats and drinks what was eaten and drunk by the peasants of ten thousand years ago; and the house he lives in has not altered as much in a thousand centuries as the fashion of a lady's bonnet in a score of weeks. But when he goes out to slay, he carries a marvel of mechanism that lets loose at the touch of his finger all the hidden molecular energies, and leaves the javelin, the arrow, the blowpipe of his fathers far behind. In the arts of peace Man is a bungler. I have seen his cotton factories and the like, with machinery that a greedy dog could have invented if it had wanted money instead of food. I know his clumsy typewriters and bungling locomotives and tedious bicycles: they are toys compared to the Maxim gun, the submarine torpedo boat. There is nothing in Man's industrial machinery but his greed and sloth: his heart is in his weapons. This marvellous force of Life of which you boast is a force of Death: Man measures his strength by his destructiveness. What is his law? An excuse for hanging you. What is his morality? Gentility! An excuse for consuming without producing. What is his art? An excuse for gloating over pictures of slaughter. What are his politics? Either the worship of a despot because a despot can kill, or parliamentary cockfighting.

DON JUAN. Pshaw! all this is old. Your weak side, my diabolic friend, is that you have always been a gull: you take Man at his own valuation. Nothing would flatter him more than your opinion of him. He loves to think of himself as bold and bad. He is neither one nor the other: he is only a coward. Call him tyrant, murderer, pirate, bully; and he will adore you, and swagger about with the consciousness of having the blood of the old sea kings in his veins. Call him liar and thief; and he will only take an action against you for libel. But call him coward; and he will go mad with rage: he will face death to outface that stinging truth. Man gives every reason for his conduct save one, every excuse for his crimes save one, every plea for his safety save one; and that one is his cowardice. Yet all his civilization is founded on his cowardice, on his abject tameness, which he calls his respectability. There are limits to what a mule or an ass will stand; but Man will suffer himself to be degraded until his vileness becomes so loathsome to his oppressors that they themselves are forced to reform it.

THE DEVIL. Precisely. And these are the creatures in whom you discover what you call a Life Force!

DON JUAN. Yes; for now comes the most surprising part of the whole business.
THE STATUE. Whats that?
DON JUAN. Why, that you can make any of these cowards brave by simply putting an idea into his head.
THE STATUE. Stuff! As an old soldier I admit the cowardice: it's as uni-

versal as sea sickness, and matters just as little. But that about putting an idea into a man's head is stuff and nonsense. In a battle all you need to make you fight is a little hot blood and the knowledge that it's more dangerous to lose than to win.

DON JUAN. That is perhaps why battles are so useless. But men never really overcome fear until they imagine they are fighting to further a universal purpose--fighting for an idea, as they call it. Why was the Crusader braver than the pirate? Because he fought, not for himself, but for the Cross. What force was it that met him with a valor as reckless as his own? The force of men who fought, not for themselves, but for Islam. They took Spain from us though we were fighting for our very hearths and homes; but when we, too, fought for that mighty idea, a Catholic Church, we swept them back to Africa.

DON JUAN. I am giving you examples of the fact that this creature Man, who in his own selfish affairs is a coward to the backbone, will fight for an idea like a hero. He may be abject as a citizen; but he is dangerous as a fanatic. He can only be enslaved whilst he is spiritually weak enough to listen to reason. I tell you, gentlemen, if you can shew a man a piece of what he now calls God's work to do, and what he will later on call by many new names, you can make him entirely reckless of the consequences to himself personally.

ANA. Yes: he shirks all his responsibilities, and leaves his wife to grapple with them.

THE STATUE. Well said, daughter. Do not let him talk you out of your common sense.

THE DEVIL. Alas! Senor Commander, now that we have got on to the subject of Woman, he will talk more than ever. However, I confess it is for me the one supremely interesting subject.

DON JUAN. To a woman, Senora, man's duties and responsibilities begin and end with the task of getting bread for her children. To her, Man is only a means to the end of getting children and rearing them.

ANA. Is that your idea of a woman's mind? I call it cynical and disgusting animalism.

DON JUAN. Pardon me, Ana: I said nothing about a woman's whole mind. I spoke of her view of Man as a separate sex. It is no more cynical than her view of herself as above all things a Mother. Sexually, Woman is Nature's contrivance for perpetuating its highest achievement. Sexually, Man is Woman's contrivance for fulfilling Nature's behest in the most economical way. She knows by instinct that far back in the evolutionary process she invented him, differentiated him, created him in order to produce something better than the single-sexed process can produce. Whilst he fulfils the purpose for which she made him, he is welcome to his dreams, his follies, his ideals, his heroisms, provided that the keystone of them all is the worship of woman, of motherhood, of the family, of the hearth. But how rash and dangerous it was to invent a separate creature whose sole function was her own impregnation! For mark what has happened. First Man has multiplied on her hands until there are as many men as women; so that she has been unable to employ for her purposes more than a fraction of the immense energy she has left at his disposal by saving him the exhausting labor of gestation. This superfluous energy has gone to his brain and to his muscle. He has become too strong to be controlled by her bodily, and too imaginative and mentally vigorous to be content with mere self-reproduction. He has created civilization without consulting her, taking her domestic labor for granted as the foundation of it.

ANA. That is true, at all events.

THE DEVIL. Yes; and this civilization! what is it, after all?

DON JUAN. After all, an excellent peg to hang your cynical commonplaces on;

but before all, it is an attempt on Man's part to make himself something more than the mere instrument of Woman's purpose. So far, the result of Life's continual effort not only to maintain itself, but to achieve higher and higher organization and completer self-consciousness, is only, at best, a doubtful campaign between its forces and those of Death and Degeneration. The battles in this campaign are mere blunders, mostly won, like actual military battles, in spite of the commanders.

THE STATUE. That is a dig at me. No matter: go on, go on.

*

At the same time that George Bernard Shaw was proclaiming an evolutionary Book of Genesis, many firm believers in the original version were calling for the legal abolishment of the teaching of evolutionary theory. Though the ranks of the fundamentalists have thinned in recent decades, they still are legislatively forceful (see "Eden and Evolution"). The classic age of conflict has passed, however; it peaked with the trial of a Tennessee schoolteacher.

On March 21, 1925, the Tennessee Legislature passed this bill:

"Be it Enacted, by the General Assembly of the State of Tennessee, that it shall be unlawful for any teacher in any of the universities, normals, and all other public schools in the State, which are supported in whole or in part by the public school funds of the State, to teach the theory that denies the story of the divine creation of man as taught in the Bible, and to teach instead that man has descended from a lower order or animals."

In order to test the constitutionality of the bill, John Scopes admitted to teaching evolution in his high school biology class. Scopes purposely had himself indicted.

William Jennings Bryan, a prominant lawyer and national political figure, immediately offered his services to the prosecution. Shortly thereafter, Clarence Darrow, criminal lawyer and writer, came to the aid of the defense. The trial proceeded with great fanfare, amidst soft drink and sandwich stands, in the summer heat of Dayton, Tennessee.

In the course of the trial, a personal battle between Bryan and Darrow soon over-shadowed the determination of the guilt or innocence of John Scopes. The debate that held the nation spellbound for eight days concerned the interpretation of the Old Testament.

As the trial drew to a close, Darrow suprised everyone by calling Bryan as a witness for the defense. The following is the closing portion of the testimony.

BRYAN AND DARROW AT DAYTON

Leslie H. Allen

Clarence Darrow

John T. Scopes as he stood before the Judges Stand and was sentenced 1925.

William J. Bryan

MR. DARROW--Mr. Bryan, do you believe that the first woman was Eve?
MR. BRYAN--Yes.
MR. DARROW--Do you believe she was literally made out of Adam's rib?
MR. BRYAN--I do.
MR. DARROW--Did you ever discover where Cain got his wife?[1]
MR. BRYAN--No, sir; I leave the agnostics to hunt for her.
MR. DARROW--You have never found out?
MR. BRYAN--I have never tried to find.
MR. DARROW--You have never tried to find?
MR. BRYAN--No.
MR. DARROW--The Bible says he got one, doesn't it? Were there other people on the earth at that time?
MR. BRYAN--I cannot say.
MR. DARROW--You cannot say? Did that never enter into your consideration?
MR. BRYAN--Never bothered me.
MR. DARROW--There were no others recorded, but Cain got a wife. That is what the Bible says. Where she came from, you don't know. All right. Does the statement, 'The morning and the evening were the first day' and 'The morning and the evening were the second day' mean anything to you?[2]
MR. BRYAN--I do not think it means necessarily a twenty-four-hour day.
MR. DARROW--You do not?
MR. BRYAN--No.
MR. DARROW--What do you consider it to be?
MR. BRYAN--I have not attempted to explain it. If you will take the second chapter--let me have the book. [Examining Bible] The fourth verse of the second chapter (Genesis) says: "These are the generations of the heavens and of the earth, when they were created, in the day that the Lord God made the earth and the heavens." The word "day" there in the very next chapter is used to describe a period. I do not see that there is necessity for construing the words, "the evening and the morning" as meaning necessarily a twenty-four hour day: "in the day when the Lord made the Heaven and the earth."
MR. DARROW--Then when the Bible said, for instance, "And God called the firmament Heaven. And the evening and the morning were the second day,"--that does not necessarily mean twenty-four hours?
MR. BRYAN--I do not think it necessarily does.
MR. DARROW--Do you think it does or does not?
MR. BRYAN--I know a great many think so.
MR. DARROW--What do you think?
MR. BRYAN--I do not think it does.
MR. DARROW--You think these were not literal days?
MR. BRYAN--I do not think they were twenty-four-hour days.
MR. DARROW--What do you think about it?
MR. BRYAN--That is my opinion--I do not know that my opinion is better on that subject than those who think it does.
MR. DARROW--Do you not think that?
MR. BRYAN--No. But I think it would be just as easy for the kind of God we believe in to make the earth in six days as in six years or in six million years or in six hundred million years. I do not think it important whether we believe one or the other.

---

[1] Genesis 4:17.
[2] Genesis 1:5, 8.

Reprinted from the transcript of the Scopes trial.

MR. DARROW--Do you think those were literal days?

MR. BRYAN--My impression is they were periods, but I would not attempt to argue as against anybody who wanted to believe in literal days.

MR. DARROW--Have you any idea of the length of the periods?

MR. BRYAN--No, I don't.

MR. DARROW--Do you think the sun was made on the fourth day?

MR. BRYAN--Yes.

MR. DARROW--And they had evening and morning without the sun?

MR. BRYAN--I am simply saying it is a period.

MR. DARROW--They had evening and morning for four periods without the sun, do you think?

MR. BRYAN--I believe in creation, as there told, and if I am not able to explain it, I will accept it.

MR. DARROW--Then you can explain it to suit yourself. And they had the evening and the morning before that time for three days or three periods. All right, that settles it. Now if you call those periods, they may have been a very long time?

MR. BRYAN--They might have been.

MR. DARROW--The creation might have been going on for a very long time?

MR. BRYAN--It might have continued for millions of years.

MR. DARROW--Yes, all right. Do you believe the story of the temptation of Eve by the serpent?

MR. BRYAN--I do.

MR. DARROW--Do you believe that after Eve ate the apple, or gave it to Adam--whichever way it was--God cursed Eve, and at that time decreed that all womankind thenceforth and forever should suffer the pains of childbirth in the reproduction of the earth?[1]

MR. BRYAN--I believe what it says, and I believe the fact as fully.

MR. DARROW--That is what it says, doesn't it?

MR. BRYAN--Yes.

MR. DARROW--And for that reason, every woman born of woman, who has to carry on the race,--the reason they have childbirth pains is because Eve tempted Adam in the Garden of Eden?

MR. BRYAN--I will believe just what the Bible says. I ask to put that in the language of the Bible, for I prefer that to your language. Read the Bible, and I will answer.

MR. DARROW--All right, I will do that: "And I will put enmity between thee and the woman."[2] That is referring to the serpent?

MR. BRYAN--The serpent.

MR. DARROW--(reading)--"And between thy seed and her seed; it shall bruise thy head, and thou shalt bruise his heel. Unto the woman He said, I will greatly multiply thy sorrow and thy conception; in sorrow thou shalt bring forth children; and thy desire shall be to thy husband, and he shall rule over thee." That is right, is it?

MR. BRYAN--I accept it as it is.

MR. DARROW--Do you believe that was because Eve tempted Adam to eat the fruit?

MR. BRYAN--I believe it was just what the Bible said.

MR. DARROW--And you believe that is the reason that God made the serpent to go on his belly after he tempted Eve?

---

[1] Genesis 3.
[2] Genesis 3: 15, 16.

MR. BRYAN--I believe the Bible as it is, and I do not permit you to put your language in the place of the language of the Almighty. You read that Bible and ask me questions, and I will answer them. I will not answer your questions in your language.

MR. DARROW--I will read it to you from the Bible: "And the Lord God said unto the serpent, Because thou hast done this, thou art cursed above all cattle, and above every beast of the field; upon thy belly shalt thou go, and dust shalt thou eat all the days of thy life." Do you think that is why the serpent is compelled to crawl upon its belly?[1]

MR. BRYAN--I believe that.

MR. DARROW--Have you any idea how the snake went before that time?

MR. BRYAN--No, sir.

MR. DARROW--Do you know whether he walked on his tail or not?

MR. BRYAN--No, sir. I have no way to know. [Laughter]

MR. DARROW--Now, you refer to the bow that was put in the heaven after the flood, the rainbow. Do you believe in that?

MR. BRYAN--Read it.

MR. DARROW--All right, Mr. Bryan, I will read it for you.

MR. BRYAN--Your Honor, I think I can shorten this testimony. The only purpose Mr. Darrow has is to slur at the Bible, but I will answer his questions. I will answer it all at once, and I have no objection in the world. I want the world to know that this man, who does not believe in a God, is trying to use a court in Tennessee-----

MR. DARROW--I object to that.

MR. BRYAN--To slur at it, and, while it will require time, I am willing to take it.

MR. DARROW--I object to your statement. I am examining you on your fool ideas that no intelligent Christian on earth believes.

---

[1] Genesis 3:14.

THE LOG FROM THE SEA OF CORTEZ
John Steinbeck

Wearing an Old Skipper's Cap, John Steinbeck strolls around his Sag Harbor home with his famous French poodle, "Charley." The pooch is the hero of the author's "Travels with Charley."

John Steinbeck was a romantic in his approach to science. In 1940, Steinbeck (a struggling author, who was to become a Nobel laureate) took a trip to the Sea of Cortez with his friend, a biologist, Ed Ricketts. Steinbeck recorded what they saw and did in a beautiful statement of ecology and natural history. The following excerpts touch briefly through-out the book -- from his original encounter with Ricketts to the start of the trip back home.

The statistics on Ed Ricketts would read: Born in Chicago, played in the streets, went to public school, studied biology at the University of Chicago. Opened a small commercial laboratory in Pacific Grove, California. Moved to Cannery Row in Monterey. Degrees--Bachelor of Science only; clubs, none; honors, none. Army service--both World Wars. Killed by a train at the age of fifty-two. Within that frame he went a long way and burned a deep scar.

I was sitting in a dentist's waiting room in New Monterey, hoping the dentist had died. I had a badly aching tooth and not enough money to have a good job done on it. My main hope was that the dentist could stop the ache without charging too much and without finding too many other things wrong.

The door to the slaughterhouse opened and a slight man with a beard came out. I didn't look at him closely because of what he held in his hand, a bloody molar with a surprisingly large piece of jawbone sticking to it. He was cursing gently as he came through the door. He held the reeking relic out to me and said, "Look at that

---

From *The Log From the Sea of Cortez*, © 1951 by John Steinbeck, reprinted by permission of The Viking Press, New York.

god-damned thing." I was already looking at it. "That came out of me," he said.

"Seems to be more jaw than tooth," I said.

"He got impatient, I guess. I'm Ed Ricketts."

"I'm John Steinbeck. Does it hurt?"

"Not much. I've heard of you."

"I've heard of you, too. Let's have a drink."

Very many conclusions Ed and I worked out together through endless discussion and reading and observation and experiment. We worked together, and so closely that I do not now know in some cases who started which line of speculation since the end thought was the product of both minds. I do not know whose thought it was.

We had a game which we playfully called speculative metaphysics. It was a sport consisting of lopping off a piece of observed reality and letting it move up through the speculative process like a tree growing tall and bushy. We observed with pleasure how the branches of thought grew away from the trunk of external reality. We believed, as we must, that the laws of thought parallel the laws of things. In our game there was no stricture of rightness. It was an enjoyable exercise on the instruments of our minds, improvisations and variations on a theme, and it gave the same delight and interest that discovered music does. No one can say, "This music is the only music" nor would we say, "This thought is the only thought," but rather, "This is a thought, perhaps well or ill formed, but a thought which is a real thing in nature."

Once a theme was established we subjected observable nature to it. The following is an example of our game--one developed quite a long time ago.

We thought that perhaps our species thrives best and most creatively in a state of semi-anarchy, governed by loose rules and half-practiced mores. To this we added the premise that over-integration in human groups might parallel the law in paleontology that over-armor or over-ornamentation are symptoms of decay and disappearance. Indeed, we thought, over-integration might be the sympton of human decay. We thought: there is no creative unit in the human save the individual working alone. In pure creativeness, in art, in music, in mathematics, there are no true collaborations. The creative principle is a lonely and an individual matter. Groups can correlate, investigate, and build, but we could not think of any group that has ever created or invented anything. Indeed, the first impulse of the group seems to be to destroy the creation and the creator. But integration, or the designed group, seems to be highly vulnerable.

Now with this structure of speculation we would slip examples on the squares of the speculative graphing paper.

Consider, we would say, the Third Reich or the Politburo-controlled Soviet. The sudden removal of twenty-five key men from either system could cripple it so thoroughly that it would take a long time to recover, if it ever could. To preserve itself in safety such a system must destroy or remove all opposition as a danger to itself. But opposition is creative and restriction is non-creative. The force that feeds growth is therefore cut off. Now, the tendency to integration must constantly increase. And this process of integration must destroy all tendencies toward improvisation, must destroy the habit of creation, since this is sand in the bearings of the system. The system then must, if our speculation is accurate, grind to a slow and heavy stop. Thought and art must be forced to disappear and a weighty traditionalism take its place. Thus we would play with thinking. A too greatly integrated system or society is in danger of destruction since the removal of one unit may cripple the whole.

Consider the blundering anarchic system of the United States, the stupidity of some of its lawmakers, the violent reaction, the slowness of its ability to change. Twenty-five key men destroyed could make the Soviet Union stagger, but we could lose our congress, our president, and our general staff and nothing much would have happened. We would go right on. In fact we might be better for it.

That is an example of the game we played. Always our thinking was prefaced with, "It might be so!" Often a whole night would draw down to a moment while we pursued the fireflies of our thinking.

We made a trip into the Gulf; sometimes we dignified it by calling it an expedition. Once it was called the Sea of Cortez, and that is a better-sounding and a more exciting name. We stopped in many little harbors and near barren coasts to collect and preserve the marine invertebrates of the littoral. One of the reasons we gave ourselves for this trip--and when we used this reason, we called the trip an expedition--was to observe the distribution of invertebrates, to see and to record their kinds and numbers, how they lived together, what they ate, and how they reproduced. That plan was simple, straight-forward, and only a part of the truth. But we did tell the truth to ourselves. We were curious. Our curiosity was not limited, but was as wide and horizonless as that of Darwin or Agassiz or Linnaeus or Pliny. We wanted to see everything our eyes would accommodate, to think what we could, and, out of our seeing and thinking, to build some kind of structure in modeled imitation of the observed reality. We knew that what we would see and record and construct would be warped, first, by the collective pressure and stream of our time and race, second by the thrust of our individual personalities. But knowing this, we might not fall into too many holes--we might maintain some balance between our warp and the separate things, the external reality. The oneness of these two might take its contribution from both. For example: the Mexican sierra has "XVII-15-IX" spines in the dorsal fin. These can easily be counted. But if the sierra strikes hard on the line so that our hands are burned, if the fish sounds and nearly escapes and finally comes in over the rail, his colors pulsing and his tail beating the air, a whole new relational externality has come into being-- an entity which is more than the sum of the fish plus the fisherman. The only way

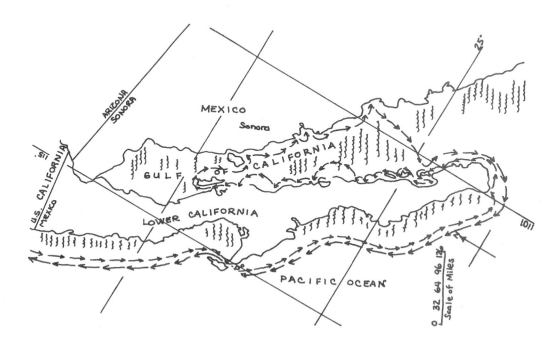

to count the spines of the sierra unaffected by this second relational reality is to sit in a laboratory, open an evil-smelling jar, remove a stiff colorless fish from formalin solution, count the spines, and write the truth "D.XVII-15-IX." There you have recorded a reality which cannot be assailed--probably the least important reality concerning either the fish or yourself.

It is good to know what you are doing. The man with his pickled fish has set down one truth and has recorded in his experience many lies. The fish is not that color, that texture, that dead, nor does he smell that way.

Such things we had considered in the months of planning our expedition and we were determined not to let a passion for unassailable little truths draw in the horizons and crowd the sky down on us. We knew that what seemed to us true could be only relatively true anyway. There is no other kind of observation. The man with his pickled fish has sacrificed a great observation about himself, the fish, and the focal point, which is his thought on both the sierra and himself.

We suppose this was the mental provisioning of our expedition. We said, "Let's go wide open. Let's see what we see, record what we find, and not fool ourselves with conventional scientific strictures. We could not observe a completely objective Sea of Cortez anyway, for in that lonely and uninhabited Gulf our boat and ourselves would change it the moment we entered. By going there, we would bring a new factor to the Gulf. Let us consider that factor and not be betrayed by this myth of permanent objective reality. If it exists at all, it is only available in pickled tatters or in distorted flashes. Let us go," we said, "into the Sea of Cortez, realizing that we become forever a part of it; that our rubber boots slogging through a flat of eelgrass, that the rocks we turn over in a tide pool, make us truly and permanently a factor in the ecology of the region. We shall take something away from it, but we shall leave something too." And if we seem a small factor in a huge pattern, nevertheless it is of relative importance. We take a tiny colony of soft corals from a rock in a little water world. And that isn't terribly important to the tide pool. Fifty miles away the Japanese shrimp boats are drudging with overlapping scoops, bringing up tons of shrimps, rapidly destroying the species so that it may never come back, and with the species destroying the ecological balance of the whole region. That isn't very important in the world. And thousands of miles away the great bombs are falling and the stars are not moved thereby. None of it is important or all of it is.

We determined to go doubly open so that in the end we could, if we wished, describe the sierra thus: "D.XVII-15-IX; A. II-15-IX," but also we could see the fish alive and swimming, feel it plunge against the lines, drag it threshing over the rail, and even finally eat it. And there is no reason why either approach should be inaccurate. Spine-count description need not suffer because another approach is also used. Perhaps out of the two approaches, we thought, there might emerge a picture more complete and even more accurate than either alone could produce. And so we went.

About noon we sailed and moved out of the shrouded and quiet Amortajada Bay and up the coast toward Marcial Reef, which was marked as our next collecting station. We arrived in mid-afternoon and collected on the late tide, on a northerly pile of boulders, part of the central reef. This was just south of Marcial Point, which marks the southern limit of Agua Verde Bay.

It was not a good collecting tide, although it should have been according to the tide chart. The water did not go low enough for exhaustive collecting. There were a few polyclads which here were high on the rocks. We found two large and many small chitons--the first time we had discovered them in numbers. There were many urchins visible but too deep below the surface to get to. Swarms of larval shrimps were in the water swimming about in small circles. The collecting was not successful in point of view of numbers of forms taken.

That night we rigged a lamp over the side, shaded it with a paper cone, and hung it close down to the water so that the light was reflected downward. Pelagic isopods and mysids immediately swarmed to the illuminated circle until the water seemed to heave and whirl with them. The small fish came to this horde of food, and on the outer edges of the light ring large fishes flashed in and out after the small fishes. Occasionally we interrupted this mad dance with dip-nets, dropping the catch into porcelain pans for closer study, and out of the nets came animals small or transparent that we had not noticed in the sea at all.

Having had no good tide at Marcial Reef, we arose at four o'clock the following morning and went in the darkness to collect again. We carried big seven-cell focusing flashlights. In some ways they make collecting in the dark, in a small area at least, more interesting than daytime collecting, for they limit the range of observation so that in the narrowed field one is likely to notice more detail. There is a second reason for our preference for night collecting--a number of animals are more active at night than in the daytime and they seem to be not much disturbed or frightened by artificial light. This time we had a very fair tide. The light fell on a monster highly colored spiny lobster in a crevice of the reef. He was blue and orange and spotted with brown. The taking of him required caution, for these big lobsters are very strong and are so armed with spikes and points that in struggling with one the hands can be badly cut. We approached with care, bent slowly down, and then with two hands grabbed him about the middle of the body. And there was no struggle whatever. He was either sick or lazy or hurt by the surf, and did not fight at all.

The cavities in Marcial Reef held a great many club-spined urchins and a number of the sharp-spined purple ones which had hurt us before. There were numbers of sea-fans, two of the usual starfish and a new species[1] which later we were to find common farther north in the Gulf. We took a good quantity of the many-rayed sun-stars, and a flat kind of cucumber which was new to us.[2] This was the first time we had collected at night, and under our lights we saw the puffer fish lazily feeding near the surface in the clear water. On the bottom, the brittle-stars, which we had always found under rocks, were crawling about like thousands of little snakes. They rarely move about in the daylight. Wherever the sharp, powerful rays of the flashlight cut into the water we could see the moving beautiful fish and the bottoms alive with busy feeding invertebrates. But collecting with a flashlight is difficult unless it is arranged that two people work together--one to hold the light and the other to take the animals. Also, from constant wetting in salt water the life of a flashlight is very short.

The one huge and beautiful lobster was the prize of this trip. We tried to photograph him on color film and as usual something went wrong but we got a very good likeness of one end of him, which was an improvement on our previous pictures. In most of our other photographs we didn't get either end.

We took several species of chitons and a great number of tunicates. There were several turbellarian flat-worms, but these are so likely to dissolve before they preserve that we had great difficulties with them. There were in the collecting pans several species of brittle-stars, numbers of small crabs and snapping shrimps, plumularian hydroids, bivalves of a number of species, snails, and some small sea-

---
[1] _Othilia tenuispinus_.
[2] Probably _Stichopus fuscus_--the specimen has since been lost sight of.

urchins. There were worms, hermit crabs, sipunculids, and sponges. The pools too had been thick with pelagic larval shrimps, pelagic isopods--tiny crustacea similar to sow-bugs--and tiny shrimps (mysids). In this area the water seemed particularly peopled with small pelagic animals--"bugs," so the boys said. Everywhere there were bugs, flying crawling, and swimming. The shallow and warm waters of the area promoted a competitive life that was astonishing.

After breakfast we pulled up the anchor and set out again northward. The pattern of the technique of the trip had by now established itself almost as a habit with us; collecting, running to a new station, collecting again. The water was intensely blue on this run, and the fish were very many. We could see the splashing of great schools of tuna in the distance where they beat the water to spray in their millions. The swordfish leaped all about us, and someone was on the bow the whole time trying to drive a light harpoon into one, but we never could get close enough. Cast after cast fell short.

We preserved and labeled as we went, and the water was so smooth that we had no difficulty with delicate animals. If the boat rolls, retractile animals such as anemones and sipunculids are more than likely to draw into themselves and refuse to relax under the Epsom salts treatment, but this sea was as smooth as a lawn, and our wake fanned out for miles behind us.

The fish-lines on the stays snapped and jerked and we brought in skipjack, Sparky's friend of the curious name, and the Mexican sierra. This golden fish with brilliant blue spots is shaped like a trout. In size it ranges from fifteen inches to two feet, is slender and a very rapid swimmer. The sierra does not seem to travel in dense, surface-beating schools as the tuna does. Although it belongs with the mackerel-like forms, its meat is white and delicate and sweet. Simply fried in big hunks, it is the most delicious fish of all.

<u>April 13</u>   At three A.M. Pacific time we passed the light on the false cape and made our new course northward and the sky was gray and threatening and the wind increased. The Gulf was blotted out for us--the Gulf that was thought and work and sunshine and play. This new world of the Pacific took hold of us and we thought again of an unseen person on the deckhouse, some kind of symbol person--to a sailor, a ghost, a premonition, a feeling in human form.

We could not yet relate the microcosm of the Gulf with the macrocosm of the sea. As we went northward the gray waves rolled up and the <u>Western Flyer</u> stubbed her nose into them and the white spray flew over us. The day passed and a new night came and the sea grew more stern. Now we plunged like a nervous horse, and no step could be taken without a steadying hand. The galley was in confusion, for a can of olive oil had leaped from its stand and flooded the floor. On the stove, the coffee pot slipped back and forth between its bars.

Over the surface of the heaving sea the birds flew landward, zigzagging to cover themselves in the wave troughs from the wind. The man at the wheel was the lucky one, for he had a grip against the pitching. He was closest to the boat and to the rising storm. He was the receiver, but also he was the giver and his hand was on the course.

What was the shape and size and color and tone of this little expedition? We slipped into a new frame and grew to be a part of it, related in some subtle way to the reefs and beaches, related to the little animals, to the stirring waters and the warm brackish lagoons. This trip had dimension and tone. It was a thing whose boundaries seeped through itself and beyond into some time and space that was more than all the Gulf and more than all our lives. Our fingers turned over the stones and we saw life that was like our life.

On the deckhouse we held the rails for support, and the blunt nose of the boat fought into the waves and the gray-green water struck us in the face. Some creative thing had happened, a real tempest in our small teapot minds. But boiling water still produces steam, whether in a watch-glass or in a turbine. It is the same stuff--weak and dissipating or explosive, depending on its use. The shape of the trip was an integrated nucleus from which weak strings of thought stretched into every reachable reality; and a reality which reached into us through our perceptive nerve trunks. The laws of thought seemed really one with the laws of things. There was some quality of music here, perhaps not to be communicated, but sounding clear and huge in our minds. The boat plunged and shook herself, and rivers of swirling water ran down the scuppers. Below in the hold, packed in jars, were thousands of little dead animals, but we did not think of them as trophies, as things cut off from the tide pools of the Gulf, but rather as drawings, incomplete and imperfect, of how it had been there. The real picture of how it had been there and how we had been there was in our minds, bright with sun and wet with sea water and blue or burned, and the whole crusted over with exploring thought. Here was no service to science, no naming of unknown animals, but rather--we simply liked it. We like it very much. The brown Indians and the gardens of the sea, and the beer and the work, they were all one thing and we were that one thing too.

The _Western Flyer_ hunched into the great waves toward Cedros Island, the wind blew off the tops of the whitecaps, and the big guy wire, from bow to mast, took up its vibration like the low pipe on a tremendous organ. It sang its deep note into the wind.

*

ANOTHER BAD TRIP
Science News

MDA (3,4-methylenedioxyamphetamine) is a mild hallucinogen that has been classified by the Department of Health, Education and Welfare as a Schedule 1 drug. This implies that the drug has a high abuse potential, is without currently accepted medical use in treatment in the United States and lacks accepted safety for use under medical supervision. This classification has managed to dry up street supplies of MDA and force some dealers to substitute other drugs. One drug that has been sold as a substitute is PMA (paramethoxyamphetamine), a substance that has five times the hallucinogenic potency of mescaline.

Since March of this year, according to the Bureau of Narcotics and Dangerous Drugs, PMA has been associated with 10 deaths, one case of temporary blindness and numerous acute overdoses throughout the United States and Canada. Accordingly, the bureau has recommended to HEW that PMA also be banned as illegal under Schedule 1. HEW has concurred and action is being taken.

Reprinted by permission of *Science News*, © 1973 by Science Service.

BRAVE NEW WORLD REVISITED
Aldous Huxley

Aldous Huxley was one of the most prophetic writers of the 20th century. His visions touched on the political and psychological, as well as the biological, future of man. In this essay, written in 1956, he evaluated his futuristic novel, Brave New World, in light of the twenty-five years since its publication in 1931. His prophecies (particularly with respect to "mind-transforming" drugs) seem all the more poignant in the perspective of an additional eighteen years.

The most distressing thing that can happen to a prophet is to be proved wrong; the next most distressing thing is to be proved right. In the twenty-five years that have elapsed since Brave New World was written, I have undergone both these experiences. Events have proved me distressingly wrong; and events have proved me distressingly right.

Here are some of the points on which I was wrong. By the early Thirties Einstein had equated mass and energy, and there was already talk of chain reaction; but the Brave New Worlders knew nothing of nuclear fission. In the early Thirties, too, we knew all about conservation and irreplaceable resources; but their supply of metals and mineral fuel was just as copious in the seventh century After Ford as ours is today. In actual fact the raw-material situation will already be subcritical by A.F. 600 and the atom will be the principal source of industrial power. Again, the Brave New Worlders had solved the population problem and knew how to maintain a permanently favorable relationship between human numbers and natural

---

"Brave New World Revisited: Proleptic Meditations on Mother's Day, Euphoria and Pavlov's Pooch" by Aldous Huxley, as it appeared in *Esquire* Magazine, copyright © 1956 by Esquire, Inc. Reprinted by Permission of Mrs. L. Huxley.

resources. In actual fact, will our descendants achieve this happy consummation within the next six centuries? And if they do achieve it, will it be by dint of rational planning, or through the immemorial agencies of pestilence, famine and internecine warfare? It is, of course, impossible to say. The only thing we can predict with a fair measure of certainty is that humanity (if its rulers decide to refrain from collective suicide) will be traveling at vertiginous speed along one of the most dangerous and congested stretches of its history.

The Brave New Worlders produced their children in biochemical factories. But though bottled babies are not completely out of the question, it is virtually certain that our descendants will in fact remain viviparous. Mother's Day is in no danger of being replaced by Bottle Day. My prediction was made for strictly literary purposes, and not as a reasoned forecast of future history. In this matter I knew in advance that I should be proved wrong.

From biology we now pass to politics. The dictatorship described in Brave New World was global and, in its own peculiar way, benevolent. In the light of current events and developing tendencies, I sadly suspect that in this forecast, too, I may have been wrong. True, the seventh century After Ford is still a long way off, and it is possible that, by then, hard economic necessity, or the social chaos resulting from nuclear warfare, or military conquest by one Great Power, or some grisly combination of all three will have bludgeoned our descendants into doing what we ought to be doing now, from motives of enlightened self-interest and common humanity--namely, to collaborate for the common good. In time of peace, and when things are going tolerably well, people cannot be expected to vote for measures which, though ultimately beneficial, may be expected to have certain disagreeable consequences in the short run. Divisive forces are more powerful than those which make for union. Vested interests in languages, philosophies of life, table manners, sexual habits, political, ecclesiastical and economic organizations are sufficiently powerful to block all attempts, by rational and peaceful methods, to unite mankind for its own good. And then there is nationalism. With its Fifty-Seven Varieties of tribal gods, nationalism is the religion of the twentieth century. We may be Christians, Jews, Moslems, Hindus, Buddhists, Confucians or Atheists; but the fact remains that there is only one faith for which large masses of us are prepared to die and kill, and that faith is nationalism. That nationalism will remain the dominant religion of the human race for the next two or three centuries at the very least seems all too probable. If total, nuclear war should be avoided, we may expect to see, not the rise of a single world state, but the continuance, in worsening conditions, of the present system, under which national states compete for markets and raw materials and prepare for partial wars. Most of these states will probably be dictatorships. Inevitably so; for the increasing pressure of population upon resources will make domestic conditions more difficult and international competition more intense. To prevent economic breakdown and to repress popular discontent, the governments of hungry countries will be tempted to enforce ever-stricter controls. Furthermore, chronic undernourishment reduces physical energy and disturbs the mind. Hunger and self-government are incompatible. Even where the average diet provides three thousand calories a day, it is hard enough to make democracy work. In a society in which most members are living on seventeen hundred to two thousand calories a day, it is simply impossible. The undernourished majority will always be ruled, from above, by the well-fed few. As population increases (twenty-seven hundred million of us are now adding to our numbers at the rate of forty millions a year, and this increase is increasing according to the rules of compound interest); as geometrically increasing demands press more and more heavily on static or, at best, arithmetically increasing supplies; as standards of living are forced down and popular discontent is forced up; as the general scramble for diminishing resources becomes ever fiercer, these national dictatorships will tend to become more oppressive at home, more ruthlessly competitive abroad. "Government," says one of the Brave New Worlders, "is an affair of sitting, not hitting. You rule with the brains and the buttocks, not the fists." But where there are many competing national dictatorships, each in trouble at home and each preparing for total or partial war against its neighbors, hitting tends to be preferred to sitting, fists, as an instrument of policy, to brains and the "masterly inactivity" (to cite Lord Salisbury's immortal phrase) of the hindquarters. In politics, the new future is likely to be closer to George Orwell's 1984 than to Brave New World.

Let me now consider a few of the points on which, I fear, I may have been right. The Brave New Worlders were the heirs and exploiters of a new kind of revolution, and this revolution was, in effect, the theme of my fable. Past revolutions have all been in fields external to the individual as a psychophysical organism—in the field, for example, of ecclesiastical organization and religious dogma, in the field of economics, in the field of political organization, in the field of technology. The coming revolution—the revolution whose consequences are described in Brave New World—will affect men and women, not peripherally, but at the very core of their organic being. The older revolutionaries sought to change the social environment in the hope (if they were idealists and not mere power seekers) of changing human nature. The coming revolutionaries will make their assault directly on human nature as they find it, in the minds and bodies of their victims or, if you prefer, their beneficiaries.

Among the Brave New Worlders, the control of human nature was achieved by eugenic and dysgenic breeding, by systematic conditioning during infancy and, later on, by "hypnopaedia," or instruction during sleep. Infant conditioning is as old as Pavlov and hypnopaedia, though rudimentary, is already a well-established technique. Phonographs with built-in clocks, which turn them on and off at regular intervals during the night, are already on the market and are being used by students of foreign languages, by actors in a hurry to memorize their parts, by parents desirous of curing their children of bed-wetting and other troublesome habits, by self-helpers seeking moral and physical improvement through autosuggestion and "affirmations of positive thought." That the principles of selective breeding, infant conditioning and hypnopaedia have not yet been applied by governments is due, in the democratic countries, to the lingering, liberal conviction that persons do not exist for the state, but that the state exists for the good of persons; and in the totalitarian countries to what may be called revolutionary conservatism—attachment to yesterday's revolution instead of the revolution of tomorrow. There is, however, no reason for complacently believing that this revolutionary conservatism will persist indefinitely. In totalitarian hands, applied psychology is already achieving notable results. One third of all the American prisoners captured in Korea succumbed, at least partially, to Chinese brainwashing, which broke down the convictions, installed by their education and childhood conditioning, and replaced these comforting axioms by doubt, anxiety and a chronic sense of guilt. This was achieved by thoroughly old-fashioned procedures, which combined straightforward instruction with what may be called conventional psychotherapy in reverse, and made no use of hypnosis, hypnopaedia or mind-modifying drugs. If all or even some of these more powerful methods had been employed, brainwashing would probably have been successful with all the prisoners, and not with a mere thirty percent of them. In their vague, rhetorical way, speech-making politicians and sermon-preaching clergymen like to say that the current struggle is not material, but spiritual—an affair not of machines, but of ideas. They forget to add that the effectiveness of ideas depends very largely on the way in which they are inculcated. A true and beneficent idea may be so ineptly taught as to be without effect on the lives of individuals and societies. Conversely, grotesque and harmful notions may be so skillfully drummed into people's heads that, filled with faith, they will rush out and move mountains—to the glory of the devil and their own destruction. At the present time the dynamism of totalitarian ideas is greater than the dynamism of liberal, democratic ideas. This is not due, of course, to the intrinsic superiority of totalitarian ideas. It is due partly to the fact that, in a world where population is fast outrunning resources, ever larger measures of governmental control become necessary—and it is easier to exercise centralized control by totalitarian than by democratic methods. Partly, too, it is due to the fact that the means employed for the dissemination of totalitarian ideas are more effective, and are used more systematically, than the means employed for disseminating democratic and liberal ideas. These more effective methods of totalitarian propaganda, education and brainwashing are, as we have seen, pretty old-fashioned. Sooner or later, however, the dictators will abandon their revolutionary conservatism and, along with it, the old-world procedures inherited from the pre-psychological and palaeo-pharmacological past. After which, heaven help us all!

Among the legacies of the proto-pharmacological past must be numbered our habit, when we feel in need of a lift, a release from tension, a mental vacation from unpleasant reality, of drinking alcohol or, if we happen to belong to a non-Western culture, of smoking hashish or opium, of chewing coca leaves or betel or any one of scores of intoxicants. The Brave New Worlders did none of these things; they merely swallowed a tablet or two of a substance called Soma. This, needless to say, was not the same as the Soma mentioned in the ancient Hindu scriptures—a rather dangerous drug derived from some as yet unidentified plant native to South Central Asia—but a synthetic, possessing "all the virtues of alcohol and Christianity, none of their defects." In small doses the Soma of the Brave New Worlders was a relaxant, an inducer of euphoria, a fosterer of friendliness and social solidarity. In medium doses it transfigured the external world and acts as a mild hallucinant; and in large doses it was a narcotic. Virtually all the Brave New Worlders thought themselves happy. This was due in part to the fact that they had been bred and conditioned to take the place assigned to them in the social hierarchy, in part to the sleep-teaching which had made them content with their lot and in part to Soma and their ability, by its means, to take holidays from unpleasant circumstances and their unpleasant selves.

All the natural narcotics, stimulants, relaxants and hallucinants known to the modern botanist and pharmacologist were discovered by primitive man and have been in use from time immemorial. One of the first things that Homo sapiens did with his newly developed rationality and self-consciousness was to set them to work finding out ways to bypass analytical thinking and to transcend or, in extreme cases, temporarily obliterate the isolating awareness of the self. Trying all things that grew in field or forest, they held fast to that which, in this context, seemed good—everything, that is to say, that would change the quality of consciousness, would make it different, no matter how, from everyday feeling, perceiving and thinking. Among the Hindus, rhythmic breathing and mental concentration have, to some extent, taken the place of the mind-transforming drugs used elsewhere. But even in the land of yoga, even among the religious and even for specifically religious purposes, cannabis indica has been freely used to supplement the effects of spiritual exercises. The habit of taking vacations from the more-or-less purgatorial world, which we have created for ourselves, is universal. Moralists may denounce it; but, in the teeth of disapproving talk and repressive legislation, the habit persists, and mind-transforming drugs are everywhere available. The Marxian formula, "Religion is the opium of the people," is reversible, and one can say, with even more truth, that "Opium is the religion of the people." In other words, mind-transformation, however induced (whether by devotional or ascetic or psycho-gymnastic or chemical means), has always been felt to be one of the highest, perhaps the very highest, of all attainable goods. Up to the present, governments have thought about the problem of mind-transforming chemicals only in terms of prohibition or, a little more realistically, of control and taxation. None, so far, has considered it in its relation to individual well-being and social stability; and very few (thank heaven!) have considered it in terms of Machiavellian statecraft. Because of vested interests and mental inertia, we persist in using alcohol as our main mind-transformer--just as our neolithic ancestors did. We know that alcohol is responsible for a high proportion of our traffic accidents, our crimes of violence, our domestic miseries; and yet we make no effort to replace this old-fashioned and extremely unsatisfactory drug by some new, less harmful and more enlightening mind-transformer. Among the Brave New Worlders, Noah's prehistoric invention of fermented liquor has been made obsolete by a modern synthetic, specifically designed to contribute to social order and the happiness of the individual, and to do so at the minimum physiological cost.

In the society described in my fable, Soma was used as an instrument of statecraft. The tyrants were benevolent, but they were still tyrants. Their subjects were not bludgeoned into obedience; they were chemically coerced to love their servitude, to cooperate willingly and even enthusiastically in the preservation of the social hierarchy. By the malignant or the ignorant, anything and everything can be used badly. Alcohol, for example, has been used, in small doses, to facilitate the exchange of thought in a symposium (literally, a drinking party) of philosophers. It has also been used, as the slave traders used it, to facilitate kidnapping. Scopolamine may be used to induce "twilight sleep"; it may also be used to increase suggestibility and soften up political prisoners. Heroin may be used to control pain; it may also be used (as it is said to have been used by the Japanese during their occupation of China) to produce an incapacitating addiction in a dangerous adversary. Directed by the wrong people, the coming revolution could be as disastrous, in its own way, as a nuclear and bacteriological war. By systematically using the psychological, chemical and electronic instruments already in existence (not to mention those new and better devices which the future holds in store), a tyrannical oligarchy could keep the majority in permanent and willing subjection. This is the prophecy I made in <u>Brave New World</u>. I hope I may be proved wrong, but am sorely afraid that I may be proved right.

Meanwhile it should be pointed out that Soma is not intrinsically evil. On the contrary, a harmless but effective mind-transforming drug might prove a major blessing. And anyhow (as history makes abundantly clear) there will never be any question of getting rid of chemical mind-transformers altogether. The choice confronting us is not a choice between Soma and nothing; it is a choice between Soma and alcohol, Soma and opium, Soma and hashish, ololiuqui, peyote, datura, agaric and all the rest of the natural mind-transformers; between Soma and such products of scientific chemistry and pharmacology as ether, chloral, veronal, benzedrine and the barbiturates. In a word, we have to choose between a more-or-less harmless all-round drug and a wide variety of more-or-less harmful and only partially effective drugs. And this choice will not be delayed until the seventh century After Ford. Pharmacology is on the march. The Soma of <u>Brave New World</u> is no longer a distant dream. Indeed, something possessing many of the characteristics of Soma is already with us. I refer to the most recent of the tranquilizing agents--the Happiness Pill, as its users affectionately call it, known in America under the trade names of Miltown and Equinel. These Happiness Pills exert a double action; they relax the tension in striped muscle and so relax the associated tensions in the mind. At the same time they act on the enzyme system of the brain in such a way as to prevent disturbances arising in the hypothalamus from interfering with the workings of the cortex. On the mental level, the effect is a blessed release from anxiety and self-regarding emotivity.

In my fable the savage expresses his belief that the advantages of Soma must be paid for by losses on the highest human levels. Perhaps he was right. The universe is not in the habit of giving us something for nothing. And yet there is a great deal to be said for a pill which enables us to assume an attitude toward circumstances of detachment, ataraxia, "holy indifference." The moral worth of an action cannot be measured exclusively in terms of intention. Hell is paved with good intentions, and we have to take some account of results. Rational and kindly behavior tends to produce good results, and these results remain good even when the behavior which produced them was itself produced by a pill. On the other hand, can we with impunity replace systematic self-discipline by a chemical? It remains to be seen.

Of all the consciousness-transforming drugs the most interesting, if not the most immediately useful, are those which, like lysergic acid and mescaline, open the door to what may be called the Other World of the mind. Many workers are already exploring the effects of these drugs, and we may be sure that other mind-transformers, with even more remarkable properties, will be produced in the near future. What man will ultimately do with these extraordinary elixirs, it is impossible to say. My own guess is that they are destined to play a part in human life at least as great as the part played, up till now, by alcohol, and incomparably more beneficent.

THE TRAGEDY OF THE COMMONS
Garrett Hardin

In this modern day classic, Garrett Hardin shows that many of our environmental crises (particularly the problem of over-population) cannot be resolved through technology or appeals to conscience. What is needed is coercion to prevent misuse of resources, and to limit population growth.

The tragedy of the commons develops in this way. Picture a pasture open to all. It is to be expected that each herdsman will try to keep as many cattle as possible on the commons. Such an arrangement may work reasonably satisfactorily for centuries because tribal wars, poaching, and disease keep the numbers of both man and beast well below the carrying capacity of the land. Finally, however, comes the day of reckoning, that is, the day when the long-desired goal of social stability becomes a reality. At this point, the inherent logic of the commons remorselessly generates tragedy.

As a rational being, each herdsman seeks to maximize his gain. Explicitly or implicitly, more or less consciously, he asks, "What is the utility to me of adding one more animal to my herd?" This utility has one negative and one positive component.

1. The positive component is a function of the increment of one animal. Since the herdsman receives all the proceeds from the sale of the additional animal, the positive utility is nearly +1.
2. The negative component is a function of the additional overgrazing created

by one more animal. Since, however the effects of overgrazing are shared by all the herdsmen, the negative utility for any particular decision-making herdsman is only a fraction of -1.

Adding together the component partial utilities, the rational herdsman concludes that the only sensible course for him to pursue is to add another animal to his herd. And another; and another...But this is the conclusion reached by each and every rational herdsman sharing a commons. Therein is the tragedy. Each man is locked into a system that compels him to increase his herd without limit--in a world that is limited. Ruin is the destination toward which all men rush, each pursuing his own best interest in a society that believes in the freedom of the commons. Freedom in a commons brings ruin to all.

Some would say that this is a platitude. Would that it were! In a sense, it was learned thousands of years ago, but natural selection favors the forces of psychological denial [8]. The individual benefits as an individual from his ability to deny the truth even though society as a whole, of which he is a part, suffers. Education can counteract the natural tendency to do the wrong thing, but the inexorable succession of generations requires that the basis for this knowledge be constantly refreshed.

A simple incident that occurred a few years ago in Leominster, Massachusetts, shows how perishable the knowledge is. During the Christmas shopping season the parking meters downtown were covered with plastic bags that bore tags reading: "Do not open until after Christmas. Free parking courtesy of the mayor and city council." In other words, facing the prospect of an increased demand for already scarce space, the city fathers reinstituted the system of the commons. (Cynically, we suspect that they gained more votes than they lost by this retrogressive act.)

In an approximate way, the logic of the commons has been understood for a long time, perhaps since the discovery of agriculture or the invention of private property in real estate. But it is understood mostly only in special cases which are not sufficiently generalized. Even at this late date, cattlemen leasing national land on the western ranges demonstrate no more than an ambivalent understanding, in constantly pressuring federal authorities to increase the head count to the point where overgrazing produces erosion and weed-dominance. Likewise, the oceans of the world continue to suffer from the survival of the philosophy of the commons. Maritime nations still respond automatically to the shibboleth of the "freedom of the seas." Professing to believe in the "inexhaustible resources of the oceans," they bring species after species of fish and whales closer to extinction [9].

The National Parks present another instance of the working out of the tragedy of the commons. At present, they are open to all, without limit. The parks themselves are limited in extent--there is only one Yosemite Valley--wheras population seems to grow without limit. The values that visitors seek in the parks are steadily eroded. Plainly, we must soon cease to treat the parks as commons or they will be of no value to anyone.

What shall we do? We have several options. We might sell them off as private property. We might keep them as public property, but allocate the right to enter them. The allocation might be on the basis of wealth, by the use of an auction system. It might be on the basis of merit, as defined by some agreed-upon standards. It might be by lottery. Or it might be on a first-come, first-served basis, administered to long queues. These, I think, are all the reasonable possibilities. They are all objectionable. But we must choose--or acquiesce in the destruction of the commons that we call our National Parks.

POLLUTION

In a reverse way, the tragedy of the commons reappears in problems of pollution. Here it is not a question of taking something out of the commons, but of putting something in--sewage, or chemical, radioactive, and heat wastes into water; noxious and dangerous fumes into the air; and distracting and unpleasnat advertising signs into the line of sight. The calculations of utility are much the same as before. The rational man finds that his share of the cost of the wastes he discharges into the commons is less than the cost of purifying his wastes before releasing them. Since this is true for everyone, we are locked into a system of "fouling our own nest," so long as we behave only as independent, rational, free-enterprisers.

The tragedy of the commons as a food basket is averted by private property, or something formally like it. But the air and waters surrounding us cannot readily be fenced, and so the tragedy of the commons as a cesspool must be prevented by different means, by coercive laws or taxing devices that make it cheaper for the polluter to treat his pollutants than to discharge them untreated. We have not progressed as far with the solution of this problem as we have with the first. Indeed, our particular concept of private property, which deters us from exhausting the positive resources of the earth, favors pollution. The owner of a factory on the bank of a stream--whose property extends to the middle of the stream--often has difficulty seeing why it is not his natural right to muddy the waters flowing past his door. The law, always behind the times, requires elaborate stitching and fitting to adapt it to this newly perceived aspect of the commons.

The pollution problem is a consequence of population. It did not much matter how a lonely American frontiersman disposed his waste. "Flowing water purifies itself every 10 miles," my grandfather used to say, and the myth was near enough to the truth when he was a boy, for there were not too many people. But as population became denser, the natural chemical and biological recycling processes became overloaded, calling for a redefinition of property rights.

HOW TO LEGISLATE TEMPERANCE?

Analysis of the pollution problem as a function of population density uncovers a not generally recognized principle of morality, namely: <u>the morality of an act is a function of the state of the system at the time it is performed</u> [10]. Using the commons as a cesspool does not harm the general public under frontier conditions, because there is no public; the same behavior in a metropolis is unbearable. A hundred and fifty years ago a plainsman could kill an American bison, cut out only the tongue for his dinner, and discard the rest of the animal. He was not in any important sense being wasteful. Today, with only a few thousand bison left, we would be appalled at such behavior.

In passing, it is worth noting that the morality of an act cannot be determined from a photograph. One does not know whether a man killing an elephant or setting fire to the grassland is harming others until one knows the total system in which his act appears. "One picture is worth a thousand words," said an ancient Chinese; but it may take 10,000 words to validate it. It is as tempting to ecologists as it is to reformers in general to try to persuade others by way of the photographic shortcut. But the essense of an argument cannot be photographed: it must be presented rationally--in words.

That morality is system-sensitive escaped the attention of most codifiers of ethics in the past. "Thou shalt not..." is the form of traditional ethical directives which make no allowance for particular circumstances. The laws of our society follow the pattern of ancient ethics, and therefore are poorly suited to governing a complex, crowded, changeable world. Our epicyclic solution is to augment statutory law with administrative law. Since it is practically impossible to spell out all the conditions under which it is safe to burn trash in the back yard or to run an automobile without smog-control, by law we delegate the details to bureaus. The result is administrative law, which is rightly feared for an ancient reason-- <u>Quis custodiet ipsos custodes?</u>--"Who shall watch the watchers themselves?" John Adams said that we must have "a government of laws and not men." Bureau administrators, trying to evaluate the morality of acts in the total system, are singularly liable to corruption, producing a government by men, not laws.

Prohibition is easy to legislate (though not necessarily to enforce); but how do we legislate temperance? Experience indicates that it can be accomplished best through the mediation of administrative law. We limit possibilities unnecessarily if we suppose that the sentiment of <u>Quis custodiet</u> denies us the use of administrative law. We should rather retain the phrase as a perpetual reminder of fearful dangers we cannot avoid. The great challenge facing us now is to invent the corrective feedbacks that are needed to keep custodians honest. We must find ways to legitimate the needed authority of both the custodians and the corrective feedbacks.

<u>FREEDOM TO BREED IS INTOLERABLE</u>

The tragedy of the commons is involved in population problems in another way. In a world governed solely by the principle of "dog eat dog"--if indeed there ever was such a world--how many children a family had would not be a matter of public concern. Parents who bred too exuberantly would leave fewer descendants, not more, because they would be unable to care adequately for their children. David Lack and others have found that such a negative feedback demonstrably controls the fecundity of birds [11]. But men are not birds, and have not acted like them for millenniums, at least.

If each human family were dependent only on its own resources; <u>if</u> the children of improvident parents starved to death; <u>if</u>, thus, overbreeding brought its own "punishment" to the germ line--<u>then</u> there would be no public interest in controlling the breeding of families. But our society is deeply committed to the welfare state [12], and hence is confronted with another aspect of the tragedy of the commons.

In a welfare state, how shall we deal with the family, the religion, the race, or the class (or indeed any distinguishable and cohesive group) that adopts overbreeding as a policy to secure its own aggrandizement [13]? To couple the concept of freedom to breed with the belief that everyone born has an equal right to the commons is to lock the world into a tragic course of action.

Unfortunately this is just the course of action that is being pursued by the United Nations. In late 1967, some 30 nations agreed to the following [14]:

> The Universal Declaration of Human Rights describes the family as the natural and fundamental unit of society. It follows that any choice and decision with regard to the size of the family must irrevocably rest with the family itself, and cannot be made by anyone else.

It is painful to have to deny categorically the validity of this right; denying it, one feels as uncomfortable as a resident of Salem, Massachusetts, who denied the reality of witches in the 17th century. At the present time, in liberal quarters, something like a taboo acts to inhibit criticism of the United Nations. There is a feeling that the United Nations is "our last and best hope," that we shouldn't find fault with it; we shouldn't play into the hands of the archconservatives. However, let us not forget what Robert Louis Stevenson said: "The truth that is suppressed by friends is the readiest weapon of the enemy." If we love the truth we must openly deny the validity of the Universal Declaration of Human Rights, even though it is promoted by the United Nations. We should also join with Kingsley Davis [15] in attempting to get Planned Parenthood-World Population to see the error of its ways in embracing the same tragic ideal.

CONSCIENCE IS SELF-ELIMINATING

It is a mistake to think that we can control the breeding of mankind in the long run by an appeal to conscience. Charles Galton Darwin made this point when he spoke on the centennial of the publication of his grandfather's great book. The argument is straightforward and Darwinian.

People vary. Confronted with appeals to limit breeding, some people will undoubtedly respond to the plea more than others. Those who have more children will produce a larger fraction of the next generation than those with more susceptible consciences. The difference will be accentuated, generation by generation.

In C. G. Darwin's words:

> It may well be that it would take hundreds of generations for the progenitive instinct to develop in this way, but if it should do so, nature would have taken her revenge, and the variety *Homo contracipiens* would become extinct and would be replaced by the variety *Homo progenitivus* [16].

The argument assumes that conscience or the desire for children (no matter which) is hereditary--but hereditary only in the most general formal sense. The result will be the same whether the attitude is transmitted through germ cells, or exosomatically, to use A. J. Lotka's term. (If one denies the latter possibility as well as the former, then what's the point of education?) The argument has here been stated in the context of the population problem, but it applies equally well to any instance in which society appeals to an individual exploiting a commons to restrain himself for the general good--by means of his conscience. To make such an appeal is to set up a selective system that works toward the elimination of conscience from the race.

PATHOGENIC EFFECTS OF CONSCIENCE

The long-term disadvantage of an appeal to conscience should be enough to condemn it; but has serious short-term disadvantages as well. If we ask a man who is exploiting a commons to desist "in the name of conscience," what are we saying to him? What does he hear?--not only at the moment but also in the wee small hours of the night when, half asleep, he remembers not merely the words we used but also the the nonverbal communication cues we gave him unawares? Sooner or later, consciously or subconsciously, he senses that he has received two communications, and that they are contradictory: (i) (intended communication "If you don't do as we ask, we will openly condemn you for not acting like a responsible citizen"; (ii) (the unintended communication) "If you do behave as we ask, we will secretly condemn you for a simpleton who can be shamed into standing aside while the rest of us exploit the commons."

Everyman then is caught in what Bateson has called a "double bind." Bateson and his co-workers have made a plausible case for viewing the double bind as an important causative factor in the genesis of schizophrenia [17]. The double bind may not always be so damaging, but it always endangers the mental health of anyone to whom it is applied. "A bad conscience," said Nietzsche, "is a kind of illness."

To conjure up a conscience in others is tempting to anyone who wishes to extend his control beyond the legal limits. Leaders at the highest level succumb to this temptation. Has any President during the past generation failed to call on labor unions to moderate voluntarily their demands for higher wages, or to steel companies to honor voluntary guidelines on prices? I can recall none. The rhetoric used on such occasions is designed to produce feelings of guilt in noncooperators.

For centuries it was assumed without proof that guilt was a valuable, perhaps even an indispensable, ingredient of the civilized life. Now, in this post-Freudian world, we doubt it.

Paul Goodman speaks from the modern point of view when he says: "No good has ever come from feeling guilty, neither intelligence, policy, nor compassion. The guilty do not pay attention to the object but only to themselves, and not even to their own interests, which might make sense, but to their anxieties" [18].

One does not have to be a professional psychiatrist to see the consequences of anxiety. We in the Western world are just emerging from a dreadful two-centuries-long Dark Ages of Eros that was sustained partly by prohibition laws, but perhaps more effectively by the anxiety-generating mechanisms of education. Alex Comfort has told the story well in The Anxiety Makers [19]; it is not a pretty one.

Since proof is difficult, we may even concede that the results of anxiety may sometimes, from certain points of view, be desirable. The larger question we should ask is whether, as a matter of policy, we should ever encourage the use of a technique the tendency (if not the intention) of which is psychologically pathogenic. We hear much talk these days of responsible parenthood; the coupled words are incorporated into the titles of some organizations devoted to birth control. Some people have proposed massive propaganda campaigns to instill responsibility into the nation's (or the world's) breeders. But what is the meaning of the responsibility in this context? Is it not merely a synonym for the word conscience? When we use the word responsibility in the absence of substantial sanctions are we not trying to browbeat a free man in a commons into acting against his own interest? Responsibility is a verbal counterfeit for a substantial quid pro quo. It is an attempt to get something for nothing.

If the word responsibility is to be used at all, I suggest that it be in the sense Charles Frankel uses it [20]. "Responsibility," says this philosopher, "is the product of definite social arrangements." Notice that Frankel calls for social arrangements--not propaganda.

MUTUAL COERCION MUTUALLY AGREED UPON

The social arrangements that produce responsibility are arrangements that create coercion, of some sort. Consider bank-robbing. The man who takes money from a bank acts as if the bank were a commons. How do we prevent such action? Certainly not by trying to control his behavior solely by a verbal appeal to his sense of responsibility. Rather than rely on propaganda we follow Frankel's lead and insist that a bank is not a commons; we seek the definite social arrangements that will keep it from becoming a commons. That we thereby infringe on the freedom of would-be robbers we neither deny nor regret.

The morality of bank-robbing is particularly easy to understand because we accept complete prohibition of this activity. We are willing to say "Thou shalt not rob banks," without providing for exceptions. But temperance also can be created by coercion. Taxing is a good coercive device. To keep downtown shoppers temperate in their use of parking space we introduce parking meters for short periods, and traffic fines for longer ones. We need not actually forbid a citizen to park as long as he wants to; we need merely make it increasingly expensive for him to do so. Not prohibition, but carefully biased options are what we offer him. A Madison Avenue man might call this persuasion; I prefer the greater candor of the word coercion.

Coercion is a dirty word to most liberals now, but it need not forever be so. As with the four-letter words, its dirtiness can be cleansed away by exposure to the light, by saying it over and over without apology or embarrassment. To many, the word coercion implies arbitrary decisions of distant and irresponsible bureaucrats; but this is not a necessary part of its meaning. The only kind of coercion I recommend is mutual coercion, mutually agreed upon by the majority of the people affected.

To say that we mutually agree to coercion is not to say that we are required to enjoy it, or even to pretend we enjoy it. Who enjoys taxes? We all grumble about them. But we accept compulsory taxes because we recognize that voluntary taxes would favor the conscienceless. We institute and (grumblingly) support taxes and other coercive devices to escape the horror of the commons.

An alternative to the commons need not be perfectly just to be preferable. With real estate and other material goods, the alternative we have chosen is the institution of private property coupled with legal inheritance. Is this system perfectly just? As a genetically trained biologist I deny that it is. It seems to me that, if there are to be differences in individual inheritance, legal possession should be perfectly correlated with biological inheritance—that those who are biologically more fit to be the custodians of property and power should legally inherit more. But genetic recombination continually makes a mockery of the doctrine of "like father, like son" implicit in our laws of legal inheritance. An idiot can inherit millions, and a trust fund can keep his estate intact. We must admit that our legal system of private property plus inheritance is unjust—but we put up with it because we are not convinced, at the moment, that anyone has invented a better system. The alternative of the commons is too horrifying to contemplate. Injustice is preferable to total ruin.

It is one of the peculiarities of the warfare between reform and the status quo that it is thoughtlessly governed by a double standard. Whenever a reform measure is proposed it is often defeated when its opponents triumphantly discover a flaw in it. As Kingsley Davis has pointed out [21], worshippers of the status quo sometimes imply that no reform is possible without unanimous agreement, an implication contrary to historical fact. As nearly as I can make out, automatic rejection of proposed reforms is based on one of two unconscious assumptions: (i) that the status quo is perfect; or (ii) that the choice we face is between reform and no action at all, while we wait for a perfect proposal.

But we can never do nothing. That which we have done for thousands of years is also action. It also produces evils. Once we are aware that the status quo is action, we can then compare its discoverable advantages and disadvantages with the predicted advantages and disadvantages of the proposed reform, discounting as best we can for our lack of experience. On the basis of such a comparison, we can make a rational decision which will not involve the unworkable assumption that only perfect systems are tolerable.

RECOGNITION OF NECESSITY

Perhaps the simplest summary of this analysis of man's population problems is this: the commons, if justifiable at all, is justifiable only under conditions of low-population density. As the human population has increased, the commons has had to be abandoned in one aspect after another.

First we abandoned the commons in food gathering, enclosing farm land and restricting pastures and hunting and fishing areas. These restrictions are still not complete throughout the world.

Somewhat later we saw that the commons as a place for waste disposal would also have to be abandoned. Restrictions on this disposal of domestic sewage are widely accepted in the Western world; we are still struggling to close the commons to pollution by automobiles, factories, insecticide sprayers, fertilizing operations, and atomic energy installations.

In a still more embryonic state is our recognition of the evils of the commons in matters of pleasure. There is almost no restriction on the propagation of sound waves in the public medium. The shopping public is assaulted with mindless music, without its consent. Our government is paying out billions of dollars to create supersonic transport which will disturb 50,000 people for every one person who is whisked from coast to coast 3 hours faster. Advertisers muddy the airwaves of radio and television and pollute the view of travelers. We are a long way from outlawing the commons in matters of pleasure. Is this because our Puritan inheritance makes us view pleasure as something of a sin, and pain (that is, the pollution of advertising) as the sign of virtue?

Every new enclosure of the commons involves the infringement of somebody's personal liberty. Infringements made in the distant past are accepted because no contemporary complains of a loss. It is the newly proposed infringements that we vigorously oppose; cries of "rights" and "freedom" fill the air. But what does "freedom" mean? When men mutually agreed to pass laws against robbing, mankind became more free, not less so. Individuals locked into the logic of the commons are free only to bring on universal ruin; once they see the necessity of mutual coercion, they become free to pursue other goals. I believe it was Hegel who said, "Freedom is the recognition of necessity."

The most important aspect of necessity that we must now recognize, is the necessity of abandoning the commons in breeding. No technical solution can rescue us from the misery of overpopulation. Freedom to breed will bring ruin to all. At the moment, to avoid hard decisions many of us are tempted to propagandize for conscience and responsible parenthood. The temptation must be resisted, because an appeal to independently acting consciences selects for the disappearance of all conscience in the long run, and an increase in anxiety in the short.

The only way we can preserve and nurture other and more precious freedoms is by relinquishing the freedom to breed, and that very soon. "Freedom is the recognition of necessity"--and it is the role of education to reveal to all the necessity of abandoning the freedom to breed. Only so, can we put an end to this aspect of the tragedy of the commons.

REFERENCES

8. G. Hardin, Ed. <u>Population, Evolution, and Birth Control</u> (Freeman, San Francisco,

1964), p. 56.
9. S. McVay, Sci. Amer. 216 (no 8):13 (1966).
10. J. Fletcher, Situation Ethics (Westminster, Philadelphia, 1966).
11. D. Lack, The Natural Regulation of Animal Numbers (Clarendon Press, Oxford, 1954).
12. H. Girvetz, From Wealth to Welfare (Stanford Univ. Press, Stanford, Calif., 1950).
13. G. Hardin, Perspec. Biol. Med 6:366 (1963).
14. U. Thant, Int. Planned Parenthood News, no. 168 (February 1968). p. 3.
15. K. Davis, Science 158:730 (1967).
16. S. Tax, Ed., Evolution after Darwin (Univ. of Chicago Press, Chicago, 1960), vol. 2, p. 469.
17. G. Bateson, D. D. Jackson, J. Haley, J. Weakland, Behav. Sci. 1:251 (1956).
18. P. Goodman, New York Rev. Books 10 (no. 8):22 (23 May 1968).
19. A. Comfort, The Anxiety Makers (Nelson, London, 1967).
20. C. Frankel, The Case for Modern Man (Harper, New York, 1955) p. 203.
21. J. D. Roslansky, Genetics and the Future of Man (Appleton-Century-Crofts, New York, 1966), p. 177.

*

SOURED ON SUGAR
Daniel Henninger

If you are finding the problems of the world to be almost more than you can endure, perhaps you can find comfort in knowing that you need not look far for something you can do to make the world a better place to live.

General Mills is test-marketing a new kids' breakfast cereal in Buffalo, N.Y., called "Mr. Wonderfull's Surprize." The packaging labeling says it is the "only cereal with a creamy chocolate flavor filling!" If it sells well in Buffalo, probably it will be marketed nationally. Mr. Wonderfull's Surprize is a box full of small globules that taste like chocolate malt balls. General Mills also has added various vitamins to Mr. Wonderfull--"at least one-third of the established minimum daily adult requirements for six vitamins and iron, plus significant amounts of two other important vitamins."

The average mother of a young child probably will greet the news of Mr. W's arrival with a big So What. She already may have learned to live with Franken Berry, Count Chocula, Cap'n Crunch, Frosted Mini-Wheats, Sugar Pops, Cocoa Krispies, Coca Pebbles, Super Sugar Crisp, Honey-Comb, Frosty O's, Frosted Rice Krinkles, and, of course, the indomitable Tony the Tiger (Frosted Flakes). So what if she now has to put up with Mr. Wonderfull and his chocolate creamy Surprize or whatever it is.

Reprinted with permission from the National Observer, copyright Dow Jones & Company, Inc. 1973.

AMBUSH IN BUFFALO

Well, there's a group of people in Buffalo who have had it up to the cavities in their lower molars with all this sweet breakfast gunk. The group, called Consumers Nutrition Monitor, is made up mostly of nutritionists, dentists, home economists, dieticians, and mothers. They are trying to drive Mr. Wonderfull's Surprize off the shelves in Buffalo before General Mills introduces it nationally.

Don Quixote surely had a better chance of conquering the windmills than these people have of stopping the Big G sugar mills. Dr. Eleanor Williams, an associate professor of nutrition at State University College at Buffalo and a leader of the down-with-Mr. Wonderfull group, concedes that about all they can do is generate publicity against the product and attempt to get across some sound nutritional advice in the process.

Nevertheless, their offensive against chocolate balls for breakfast highlights an issue that consumer groups and the makers of these cereals have been fighting over for some time.

THE PROS AND CONS

The consumer groups criticize pre-sweetened breakfast cereals for their sugar content, which they say contributes to tooth decay and predisposes children to crave other sugar-sweetened foods. They say our national sweet tooth has caused a billion-dollar dental-health problem in this country.

The cereal makers reply that most pre-sweetened cereals are now highly fortified with vitamins, a recent change that General Mills concedes was a response to consumer agitation for more-nutritious breakfast cereals. On the sugar issue the manufacturers argue that sweetness ensures that kids will eat the newly nutritious foods. As for dental disease, they say there is no scientific proof that pre-sweetened cereals cause tooth decay.

There is not much disagreement on the nutritional value of fortified breakfast cereals. Many consumerists acknowledge that some of the highly fortified, ready-to-eat cereals provide a good, healthy supply of some important vitamins and minerals. In his book <u>Nutrition Scoreboard</u>, Michael Jacobson of the Center for Science in the Public Interest highly rates Product 19, King Vitaman, Raisin Bran, Kellogg's Concentrate, Fortified Oat Flakes, Wheat Chex, Cheerios, Wheaties, Special K, and Life.

SUGAR IS THE BIG ISSUE

The tooth-decay problem in this country is pandemic, and sugar (which has no nutritional value) is one of the primary causes of it. It is estimated that we annually spend about $2 billion repairing decayed teeth. There are not enough dentists available for all the teeth that need filling; if there were, we would spend about $8 billion annually. The American Dental Association says sugar clearly causes tooth decay and points with dismay to statistics indicating that on a per capita basis we consume about two pounds of sugar <u>each week</u>.

Dr. Abraham E. Nizel of the Tufts University School of Dental Medicine repeated to a congressional committee earlier this year the dental profession's indictment of sugar (he spoke for the dental association). Though Nizel **came** down hard on cough drops, candy, chewing gum, and soft drinks, he didn't specifically mention pre-sweetened breakfast cereals. I asked him about that and he threw the cereals on the fire: "Just the fact that there is sugar present unquestionably can lead to a caries-producing situation." (Dentists call cavities "caries," a derivative of a word meaning "rotten.")

FREQUENCY IS KEY FACTOR

The cereal producers point out that the sugar in the products accounts for only about 3 per cent of a child's annual sugar intake. Nizel dismisses that defense, arguing that the _amount_ of sugar ingested isn't the problem; the _frequency_ of sugar eating causes the trouble. In simple terms this is what happens: Sugar consumption causes the formation of a tooth surface of a filmy substance called "plaque." Whenever additional sugar hits that plaque, it creates about 20 to 30 minutes of decay-causing acid.

Nizel argues that a child who eats a sweetened cereal for breakfast, then snacks through the day on soda pop, cookies, and other sweet foods causes a constant assault on his or her teeth. A date with the dentist's drill is the result. The back panel of Mr. Wonderfull's Surprize suggests serving it as a snack, citing its nutritional value. According to Nizel, snacking on a chocolate Surprize or any of the sweetened cereals contributes to the decaying action in a kid's mouth.

Personally I don't quite see the logic in the cereal makers' argument that because studies haven't specifically identified their sweetened cereals as causing tooth decay, their products have nothing to do with all those young teeth going bad. It seems to me that you don't need a million-dollar study to know that a lot of kids who get hooked on such sweeties as Mr. Wonderfull or Frosted Rice Krinkles are going to eat a lot of other sweet, less-nutritious junk. The cereal men ought to withdraw their breakfast gooeys and let the kids suffer through a little sugar withdrawal.

*

The Scientific Community treats
evolution more as a principle than
a theory. For many fundamentalists,
however, evolution is a threat to
basic religious beliefs.

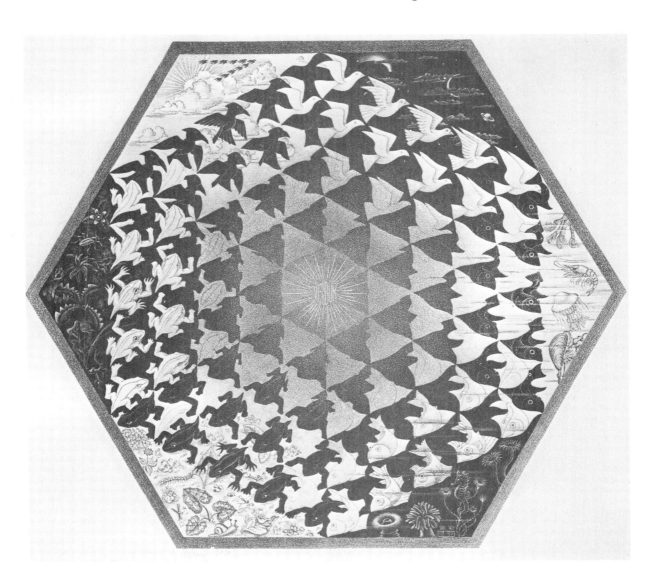

PART II
EVOLUTION

EDEN AND EVOLUTION
Ron Moskowitz

   Thousands of California children are taught at home that the Bible's version of the origin of man is the only true account. These same children are then confronted at public school with Charles Darwin's evolutionary explanation of how man came to be. For parents in home after home the conflicting explanations have caused a crisis of belief. Last month fundamentalist churchgoers, who believe that the Bible is the literal word of God, capped nearly a decade of hard lobbying with a major victory. The State Board of Education agreed to alter the wording of new science textbooks so that evolution will be taught as speculative theory, not "scientific dogma."
   From now on children in the fourth and sixth grades will learn that Darwin's theory of evolution is simply science's way of guessing how man was created. Modified textbooks will expose what the board has decided are holes in the theory--areas in which scientists have no hard proof for evolution.
   The decision is expected to have nationwide repercussions. California buys 10 per cent of all the textbooks published in the United States, and most publishers design their textbooks for this lucrative market. Once the demanded changes are made to satisfy the Board of Education in the nation's most populous state, it is unlikely that publishers will spend the additional sums necessary to publish an unexpurgated version for the other forty-nine states.

> "And God created man in his own image,
>   in the image of God
> He created him; male and female
> He created them."--Genesis 1:27

Copyright © 1973 by Saturday Review Co. First appeared in *Saturday Review*, February 1973. Used with permission.

Scientists feared the board would go even farther and give in to the fundamentalists' demand that competing religious theories about man's origin be put side by side with Darwin's. The evolutionists won a compromise only by bringing big guns to bear, including the National Academy of Sciences, the American Chemical Society, various associations representing science teachers, nineteen Nobel prize-winning scientists in California, and several faculty senates at the campuses of the University of California.

An editorial in the journal of the American Association for the Advancement of Science spelled out clearly the arguments of all of these groups. It was read at a marathon hearing the board held in November that almost rivaled the Eden-versus-Evolution debates of the 1925 Scopes trial. "During the past century and a half," the editorial stated, "the earth's crust and the fossils preserved in it have been intensively studied by geologists and paleontologists. Biologists have intensively studied the origin, structure, physiology, and genetics of living organisms. The conclusion of these studies is that the living species of animals and plants have evolved from different species that lived in the past. The scientists involved in these studies have built up the body of knowledge known as the biological theory of the origin and evolution of life. There is no currently acceptable alternative scientific theory to explain the phenomena.

"The various accounts of creation that are part of the religious heritage of many people are not scientific statements or theories. They are statements that one may choose to believe, but if he does, this is a matter of faith, because such studies are not subject to study or verification by the procedures of science. A scientific statement must be capable of test by observation and experiment. It is acceptable only if, after repeated testing, it is found to account satisfactorily for the phenomena to which it is applied."

The fundamentalists argued that there are large unexplained gaps in the fossil record. Therefore, they said, believing in evolution was just as much a matter of faith. They also charged that despite such gaps, overzealous scientists have created a sort of nontheistic religion out of evolution.

Under the terms of the ruling, changes will go into the textbooks in September 1974. In one book the statement "It is known that life began in the seas" will be altered to read: "Most scientists believe that life may have begun in the seas." Another sentence in the book states that the oldest rocks contain no fossils and adds: "These rocks are from periods before life began or from periods when the only forms of life were minute and soft-bodied and left no fossil remains." This would be changed to read: "These rocks are from periods before life began or from periods when the only forms of life left no fossil remains. Thus scientists can only speculate about the character of these early life forms."

Another book now states: "All scientists do not agree on when and how the earth was formed." This "dogma" would be changed to read: "Scientists are not sure when and how the earth was formed"--a statement sure to instill confidence in every child who reads it.

How could the Board of Education in the largest state in the union unanimously decide that the entire scientific community was wrong and a handful of laymen were right? It took nearly ten years of pressure, carried on with evangelical zeal and coupled with abrupt political changes, to evolve that situation.

In May of 1963 Mrs. Jean Sumrall and several like-minded women first appeared before the board to protest the teaching of Darwin as fact. The board at that time was markedly liberal, having been appointed by Democratic Governor Edmund G. "Pat" Brown. But in December of that year even the board made a small concession to the churchgoers, unanimously approving the following policy statement:

"Future state textbooks dealing with the subject of man's origins should refer to Darwinian evolution as an important scientific theory or hypothesis. California teachers should be encouraged to teach Darwinian evolution as theory rather than as a permanent, unchanging truth ...."

This appeased the petitioners somewhat, but they had only begun to fight. Their next target--equal billing of Eden and evolution in the books. In 1966 Governor Ronald Reagan was elected, and the political pendulum moved sharply to the Right. The creationists approached the board again, only to find that new science textbooks were not up for adoption that year. Three years later, however, when the board was considering textbook specifications, they returned.

"I cannot look at the universe as the result of blind chance, yet I can see no evidence of beneficent design, or indeed of design of any kind, in the details."--Charles Darwin

The battle widened. Although the new Board of Education was more conservative, the State Curriculum Commission was not, and it protested publicly the board's announced intention to make Darwin compete with Genesis. The commission pleaded with the board to approve the framework it had written, but the all-Christian board refused. When the textbook framework was finally approved the following year, it contained two new paragraphs calling for competing religious theories.

Of all the publishers competing for part of the multimillion-dollar contract, however, only two submitted books that attempted to meet the criteria; neither of these was recommended, because the commission felt the overall quality was too low. One went so far as to pull its biographical sketch of paleontologist L. S. B. Leakey and replace it with Michelangelo's version of the Creation. A ten-page chapter on genetics was also chucked in favor of a lengthy discussion of the various theories of the origin of man. Last month the board--which has final say on what curriculum materials are used in the first eight grades--nearly adopted these two books or others like them.

The battle in California is not over. One of the ten seats on the board is vacant. If Governor Reagan appoints another member sympathetic to the fundamentalists, the creationists will have the six votes they need and pupils in the Golden State might yet be learning about creationist theories in science classes.

At the meeting last month the creationists made it clear their campaign would end only with total victory. Asked if she was happy at their accomplishments so far, Mrs. Sumrall agreed that she was. "Now I'm going to ask for more," she said.

LAND LIFE: A SPECULATION
Isaac Asimov

Isaac Asimov is undoubtedly the most prolific science writer in the world. He has done more than anyone else to popularize biology. Asimov really deserves a larger place in an anthology of the life sciences, but, then, he can be found just about anywhere else.

As far as paleontologists can tell by careful study of carbon traces in very ancient rocks, life may have begun on Earth as long as three billion years ago. It seems to have begun in the sea; it was a mere 400 million years ago that the dry land began to be colonized by life forms.

Presumably, the colonization of land took place by way of the tides. In the twice-daily wash of water up the shore and back, living organisms of various kinds were left behind and gradually underwent adaptation enabling them to survive the temporary absence of water. Through natural selection some were able to live on dry land permanently. But why did it take so long? Why did life remain at sea for more than two and a half billion years, with dry land untenanted, and then spread out over the land surfaces?

Here is another, apparently unrelated puzzle. Why does Earth have a large moon? Small planets—Pluto, Venus, and Mercury, for example—have no satellites at all. All the other planets have satellites that are one-thousandth their own size or less. Mars has two satellites, each only a dozen miles across. The Earth, however, has the moon, which is one-eightieth its own size and rivals the largest satellites of the giant planets.

Copyright © 1973 by Saturday Review Co. First appeared in *Saturday Review*, February 1973. Used with permission of the publisher and the author.

What's more, the moon doesn't revolve in the plane of Earth's equator, as one would expect a satellite to do, but revolves more or less in the plane of the sun's equator, as a planet would do. Then, too, the moon rocks brought back by astronauts show us that the moon's crust is far more heat-baked than one would suspect at its present distance from the sun. Might it be that the moon was once an independent planet that passed fairly close to the sun at one end of its orbit? That would heat-bake it. At the other end it might have come fairly close to Earth's orbit and eventually could have been captured by Earth. There are even reasons for suggesting that the capture took place as recently as 600 million years ago.

Thus, many million years may have passed before the moon settled down in its present nearly circular, gradually retreating orbit. During that period of time the tides of Earth, thanks to a generally close moon, would have been extremely active. It may have been in this wild interval that sea life was dashed onto the shore with enough vigor--and exposed to water absence with enough pertinacity--to supply sufficient pressure for the development of land life.

It appears that the capture of a large satellite like the moon is a rare and unlikely event. But what if it is only through the capture of such a satellite, and a long following period of giant tides, that land life can develop? Then man might find himself in an odd position.

Astronomers suspect there may be uncounted millions of Earth-like planets in the universe that bear life. But of them all, how many would bear _land_ life? Might it be that in a universe swarming with life, it is only the _fish_ who have company and that we ourselves are alone after all?

AFRICAN FOSSIL BONANZA POINTS TO EARLIEST MAN
Stanley Meisler

Mankind exists within the flow of evolution, and we are finding traces of our primordial ancestors.

KOOBI FORA, Lake Rudolf, Kenya--There can be few areas on earth so difficult for man as this land by the shores of Lake Rudolf in northern Kenya. Yet man may have thrived here almost 3 million years ago.

It is strange knowing this and seeing the area now. The Gabbra people, who once lived here, were driven away in the last few years by the dearth of fresh water. They left little human life behind.

Occasionally, nomads bring their sheep and goats from Ethiopia to devour the few patches of grass amid the scrub bush in the baked land. Some fishermen work the alkaline lake, netting their food out of the algae green waters, not far from slithering, ugly crocodiles.

Between 2 and 3 million years ago, there were few areas on earth so hospitable to man. While ice menaced the rest of the planet, this land near the equator had warmth. There were large expanses of lush, swampy grass. Fresh rivers cut across the plains, their banks rich with forest. Heavy rains fell.

GREAT MINE OF PALEONTOLOGY

This part of the world was so hospitable that it has now become one of the great mines of palenotology.

Copyright 1973 Los Angeles Times. Used with permission.

No other place on earth has yielded older fossil specimens of what seems like early man. The most spectacular yield came last year. More than 150 pieces of fossil bone were found, which, when pieced together, became a skull, now classified as that of an early man 2.9 million years old.

This discovery, and many others like it, could change most of the standard scientific theories about how and when man evolved.

The first bits of this skull were spotted in August, 1972, by Bernard Ngeneo, a member of a team of eight Kenya fossil hunters working for the expedition of Richard Leakey, the 29-year-old director of the Kenya National Museum and the son of the late Dr. Louis S. B. Leakey, the renowned Kenya paleontologist.

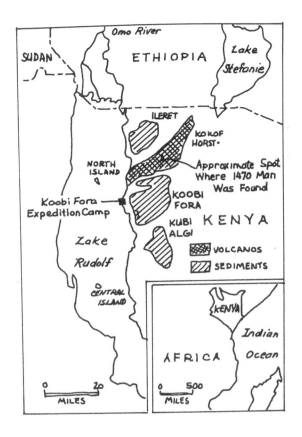

90 SPECIMENS OF EARLY MAN

Since Richard Leakey began work at Lake Rudolf in 1968, more than 90 specimens have been found there of early man or of creatures akin to man. The Leaky expedition, which will use the services of 26 other scientists this summer, has surely become science's most exciting and productive hunt these days for the fossils of early man.

The 700 square miles that make up the Lake Rudolf site are so rich in fossils that, to a layman, the job of hunting them seems deceptively simple.

On a recent Saturday, Leakey flew a couple of visitors in his single-engine plane from the expedition camp to Koobi Fora 30 miles inland to the area where the fossil skull of what may be the earliest man--known, by its museum catalogue number, simply as 1470 Man--was found.

DRUGS MAY BE UNABLE TO CONTROL SOME KILLER DISEASES, EXPERTS SAY

We can watch the process of evolution work, with miraculous (albeit, threatening) speed. Because bacteria are so rapid and prolific in their reproduction, strains which are genetically resistant to antibiotics can develop in a matter of years.

ATLANTA--Several major killer diseases have become resistant to antibiotics and are making comebacks in some areas of the world, the Center for Disease Control says.

Penicillin, ampicillin, streptomycin, tetracycline and other so-called miracle drugs that have reduced suffering and death caused by serious diseases may not be able to cope with new strains of gonorrhea, typhoid, malaria, tuberculosis and other maladies, according to doctors at the center.

Typhoid and malaria are rarely found in the United States, but gonorrhea is said by many experts to be at the epidemic stage. Tuberculosis, while rare, is a major health problem in many of the country's urban ghettos.

Doctors at the Atlanta-based center note that a new typhoid strain in Mexico is resisting treatment by drugs. Doctors in Mexico have been fighting a typhoid epidemic for more than two years, and their problems have been compounded by the drug-resistant strain.

The bacteria that causes gonorrhea has developed such resistance to penicillin that the recommended dosage now is about eight times the amount originally required.

Experts at the center continually test other disease strains to determine if any are becoming resistant to drugs.

"Thank goodness we have not seen any resistance developed in cholera or bubonic

Used by permission of Associated Press.

plague," said Dr. Eugene Gangarosa, head of the center's epidemiology branch.

The center has had to change its drug treatment recommendations several times in the last 20 years to keep ahead of resistant diseases.

The head of the center's venereal disease section, Dr. Paul Wiesner, said growing misuse of antibiotics probably was the biggest single factor contributing to the rise of drug-resistant diseases in recent years.

Wiesner said many gonorrhea patients got drugs for their ailment from a doctor and then shared the antibiotics with a sex partner.

"That way, both persons get only half the dose needed to knock out his bug," said Wiesner. "And that way two persons are in danger of their gonorrhea building up a resistance to drugs."

Experts at the center do not rule out the possibility that drug-resistant diseases could become major problems in this country.

Gongarosa said a typhoid epidemic that occurred last year in Dade County, Fla., emphasized the need for disease fighters to keep a constant vigilance on maladies around the world.

WHAT AGGRESSION IS GOOD FOR
Konrad Lorenz

Evolution affects behavioral, as well as physical, characteristics of animals. In his controversial book, On Aggression, Konrad Lorenz looked at the evolutionary value of aggression. He has not excepted humans from his conclusions, and has been roundly criticized for his view of mankind as an innately aggressive species (see "The Scarred Species"). Recently, however, Lorenz shared a Nobel prize with two fellow ethologists, Karl von Frisch and Nicholas Tinnbergan.

What is the value of all this fighting? In nature, fighting is such an ever-present process, its behavior mechanisms and weapons are so highly developed and have so obviously arisen under the selection pressure of a species-preserving function, that it is our duty to ask this Darwinian question.

The layman, misguided by sensationalism in press and film, imagines the relationship between the various "wild beasts of the jungle" to be a bloodthirsty struggle, all against all. In a widely shown film, a Bengal tiger was seen fighting with a python, and immediately afterward the python with a crocodile. With a clear conscience I can assert that such things never occur under natural conditions. What advantage would one of these animals gain from exterminating the other? Neither of them interferes with the other's vital interests.

Darwin's expression, "the struggle for existence," is sometimes erroneously interpreted as the struggle between different species. In reality, the struggle Darwin was thinking of and which drives evolution forward is the competition between

From *On Aggression* by Konrad Lorenz, copyright © 1963 by Dr. G. Borotha-Schoeler Verlag, Wien, Austria; English translation copyright © 1966 by Konrad Lorenz. Reprinted by permission of Harcourt Brace Jovanovich, Inc.

near relations. What causes a species to disappear or become transformed into a different species is the profitable "invention" that falls by chance to one or a few of its members in the everlasting gamble of hereditary change. The descendants of these lucky ones gradually outstrip all others until the particular species consists only of individuals who possess the new "invention."

There are, however, fightlike contests between members of different species: at night an owl kills and eats even well-armed birds of prey, in spite of their vigorous defense, and when these birds meet the owl by day they attack it ferociously. Almost every animal capable of self-defense, from the smallest rodent upward, fights furiously when it is cornered and has no means of escape. Besides these three particular types of inter-specific fighting, there are other, less typical cases; for instance, two cave-nesting birds of different species may fight for a nesting cavity. Something must be said here about these three types of inter-specific fighting in order to explain their peculiarity and to distinguish them from the intra-specific aggression which is really the subject of this book.

The survival value of inter-specific fights is much more evident than that of intra-specific contests. The way in which a predatory animal and its prey influence each other's evolution is a classical example of how the selection pressure of a certain function causes corresponding adaptations. The swiftness of the hunted ungulate forces its feline pursuers to evolve enormous leaping power and sharply armed toes. Paleontological discoveries have shown impressive examples of such evolutionary competition between weapons of attack and those of defense. The teeth of grazing animals have achieved better and better grinding power, while, in their parallel evolution, nutritional plants have devised means of protecting themselves against being eaten, as by the storage of silicates and the development of hard, wooden thorns. This kind of "fight" between the eater and the eaten never goes so far that the predator causes extinction of the prey: a state of equilibrium is always established between them, endurable by both species. The last lions would have died of hunger long before they had killed the last pair of antelopes or zebras; or, in terms of human commercialism, the whaling industry would go bankrupt before the last whales became extinct. What directly threatens the existence of an animal species is never the "eating enemy: but the competitor. In prehistoric times man took the Dingo, a primitive domestic dog, to Australia. It ran wild there, but it did not exterminate a single species of its quarry; instead, it destroyed the large marsupial beasts of prey which ate the same animals as it did itself. The large marsupial predators, the Tasmanian Devil and the Marsupial Wolf, were far superior to the Dingo in strength, but the hunting methods of these "old-fashioned," relatively stupid and slow creatures were inferior to those of the "modern" mammal. The Dingo reduced the marsupial population to such a degree that their methods no longer "paid," and today they exist only in Tasmania, where the Dingo has never penetrated.

In yet another respect the fight between predator and prey is not a fight in the real sense of the word: the stroke of the paw with which a lion kills his prey may resemble the movements that he makes when he strikes his rival, just as a shotgun and a rifle resemble each other outwardly; but the inner motives of the hunter are basically different from those of the fighter. The buffalo which the lion fells provokes his aggression as little as the appetizing turkey which I have just seen hanging in the larder provokes mine. The differences in these inner drives can clearly be seen in the expression movements of the animal: a dog about to catch a hunted rabbit has the same kind of excitedly happy expression as he has when he greets his master or awaits some longed-for treat. From many excellent photographs it can be seen that the lion, in the dramatic moment before he springs, is in no way angry. Growling, laying the ears back, and other well-known expression movements of fighting behavior are seen in predatory animals only when they are very afraid of a wildly resisting prey, and even then the expressions are only suggested.

The opposite process, the "counteroffensive" of the prey against the predator, is more nearly related to genuine aggression. Social animals in particular take every possible chance to attack the "eating enemy" that threatens their safety. This process is called "mobbing." Crows or other birds "mob" a cat or any other nocturnal predator, if they catch sight of it by day.

The survival value of this attack on the eating enemy is self-evident. Even if the attacker is small and defenseless, he may do his enemy considerable harm. All animals which hunt singly have a chance of success only if they take their prey by surprise. If a fox is followed through the wood by a loudly screaming jay, or a sparrow hawk is pursued by a flock of warning wagtails, his hunting is spoiled for the time being. Many birds will mob an owl, if they find one in the daytime, and drive it so far away that it will hunt somewhere else the next night. In some social animals such as jackdaws and many kinds of geese, the function of mobbing is particularly interesting. In jackdaws, its most important survival value is to teach the young, inexperienced birds what a dangerous eating enemy looks like, which they do not know instinctively. Among birds, this is a unique case of traditionally acquired knowledge.

Geese and ducks "know" by very selective, innate releasing mechanisms that anything furry, red-brown, long-shaped, and slinking is extremely dangerous, but nonetheless mobbing, with its intense excitement and the gathering together of geese from far and wide, has an essentially educational character as well as a survival value; anyone who did not know it already learns: foxes may be found <u>here!</u> At a time when only part of the shore of our lake was protected by a fox-proof fence, the geese kept ten or fifteen yards clear of all unfenced cover likely to conceal a fox, but in the fenced-in area they penetrated fearlessly into the thickets of young fir trees. Besides this didactic function, mobbing of predators by jackdaws and geese still has the basic, original one of making the enemy's life a burden. Jackdaws actively attack their enemy, and geese apparently intimidate it with their cries, their thronging, and their fearless advance. The great Canada geese will even follow a fox over land in a close phalanx, and I have never known a fox in this situation try to catch one of his tormentors. With ears laid back and a disgusted expression on his face, he glances back over his shoulder at the trumpeting flock and trots slowly--so as not to lose face--away from them.

Among the larger, more defense-minded herbivores which, en masse, are a match for even the biggest predators, mobbing is particularly effective; according to reliable reports, zebras will molest even a leopard if they catch him on a veldt where cover is sparse. The reaction of social attack against the wolf is still so ingrained in domestic cattle and pigs that one can sometimes land oneself in danger by going through a field of cows with a nervous dog which, instead of barking at them or at least fleeing independently, seeks refuge between the legs of its owner. Once, when I was out with my bitch Stasi, I was obliged to jump into a lake and swim for safety when a herd of young cattle half encircled us and advanced threateningly; and when he was in southern Hungary during the First World War my brother spent a pleasant afternoon up a tree with his Scotch terrier under his arm, because a herd of half-wild Hungarian swine, disturbed while grazing in the wood, encircled them, and with bared tusks and unmistakable intentions began to close in on them.

Much more could be said about these effective attacks on the real or supposed enemy. In some birds and fishes, to serve this special purpose brightly colored "aposematic" or warning colors have evolved, which predators notice and associate with unpleasant experiences with the particular species. Poisonous, evil-tasting, or otherwise specially protected animals have, in many cases, "chosen" for these warning signals the combination of red, white, and black; and it is remarkable that the Common Sheldrake and the Sumatra Barb, two creatures which have nothing in common either with each other or the above-named groups, should have done the same thing. It has long been known that Common Sheldrake mob predatory animals and that they so disgust the fox with the sight of their brightly colored plummage that they can nest safely in inhabited foxholes. I bought some Sumatra Barb because I had asked myself why these fishes looked so poisonous; in a large communal aquarium, they immediately answered my question by mobbing big Cichlids so persistently that I had to save the giant predators from the only apparently harmless dwarfs.

There is a third form of fighting behavior, and its survival value is as easily demonstrated as that of the predator's attack on its prey or the mobbing by the prey of the eating enemy. With H. Hediger, we call this third behavior pattern the <u>critical reaction</u>. The expression "fighting like a cornered rat" has become symbolic of the desperate struggle in which the fighter stakes his all, because he cannot escape and can expect no mercy. This most violent form of fighting behavior is motivated by fear, by the most intense flight impulses whose natural outlet is prevented by the fact that the danger is too near; so the animal, not daring to turn its back on it, fights with the proverbial courage of desperation. Such a contingency may also occur when, as with the cornered rat, flight is prevented by lack of space, or by strong social ties, like those which forbid an animal to desert its brood or family. The attack which a hen or goose makes on everything that goes too near her chicks or goslings can also be classified as a critical reaction. Many animals will attack desperately when surprised by an enemy at less than a certain critical distance, whereas they would have fled if they had noticed his coming from farther away. As Hediger has described, lion tamers maneuver their great beasts of prey into their positions in the arena by playing a dangerous game with the margin between flight distance and critical distance; and thousands of big game hunting stories testify to the dangerousness of large beasts of prey in dense cover. The reason is that in such circumstances the flight distance is particularly small, because the animal feels safe, imagining that it will not be noticed by a man even if he should penetrate the cover and get quite close; but if in so doing the man oversteps the animal's critical distance, a so-called hunting accident happens quickly and disastrously.

All the cases described above, in which animals of different species fight against each other, have one thing in common: every one of the fighters gains an obvious advantage by its behavior or, at least, in the interests of preserving the species it "ought to" gain one. But intra-specific aggression, aggression in the proper and narrower sense of the word, also fulfills a species-preserving function. Here, too, the Darwinian question "What for?" may and must be asked. Many people will not see the obvious justification for this question, and those accustomed to the classical psychoanalytical way of thinking will probably regard it as a frivolous attempt to vindicate the life-destroying principle or, purely and simply, evil. The average normal civilized human being witnesses aggression only when two of his fellow citizens or two of his domestic animals fight, and therefore sees only its evil effects. In addition there is the alarming progression of aggressive actions ranging from cocks fighting in the barnyard to dogs biting each other, boys thrashing each other, young men throwing beer mugs at each other's heads, and so on to bar-room brawls about politics, and finally to wars and atom bombs.

With humanity in its present cultural and technological situation, we have good reason to consider intra-specific aggression the greatest of all dangers. We shall not improve our chances of counteracting it if we accept it as something metaphysical and inevitable, but on the other hand, we shall perhaps succeed in finding remedies if we investigate the chain of its natural causation. Wherever man has achieved the power of voluntarily guiding a natural phenomenon in a certain direction, he has owed it to his understanding of the chain of causes which formed it. Physiology, the science concerned with the normal life processes and how they fulfill their species-preserving function, forms the essential foundation for pathology, the science investigating their disturbances. Let us forget for a moment that the aggression drive has become derailed under conditions of civilization, and let us inquire impartially into its natural causes. For the reasons already given, as good Darwinians we must inquire into the species-preserving function which, under natural--or rather precultural--conditions, is fulfilled by fights within the species, and which by the process of selection has caused the advanced development of intra-specific fighting behavior in so many higher animals. It is not only fishes that fight their own species: the majority of vertebrates do so too, man included.

Darwin had already raised the question of the survival value of fighting, and he has given us an enlightening answer: It is always favorable to the future of a species if the stronger of two rivals takes possession either of the territory or of the desired female. As so often, this truth of yesterday is not the untruth of today but only a special case; ecologists have recently demonstrated a much more essential function of aggression. Ecology--derived from the Greek oikos, the house--is the branch of biology that deals with the manifold reciprocal relations of the organism to its natural surroundings--its "household"--which of course includes all other animals and plants native to the environment. Unless the special interests of a social organization demand close aggregation of its members, it is obviously most expedient to spread the individuals of an animal species as evenly as possible over the available habitat. To use a human analogy: if, in a certain area, a larger number of doctors, builders, and mechanics want to exist, the representatives of these professions will do well to settle as far away from each other as possible.

The danger of too dense a population of an animal species settling in one part of the available biotope and exhausting all its sources of nutrition and so starving can be obviated by a mutual repulsion acting on the animals of the same species, effecting their regular spacing out, in much the same manner as electrical charges are regularly distributed all over the surface of a spherical conductor. This, in plain terms, is the most important survival value of intra-specific aggression.

Now we can understand why the sedentary coral fish in particular are so crazily colored. There are few biotopes on earth that provide so much and such varied nutrition as a coral reef. Here fish species can, in an evolutionary sense, take up very different professions: one can support itself as an "unskilled laborer," doing what any average fish can do, hunting creatures that are neither poisonous nor armor-plated nor prickly, in other words hunting all the defenseless organisms approaching the reef from the open sea, some as "plankton," others as active swimmers "intending" to settle on the reef, as millions of free-swimming larvae of all coral-dwelling organisms do. On the other hand, another fish species may specialize in eating forms of life that live on the reef itself and are therefore equipped with some sort of protective mechanism which the hunting fish must render harmless. Corals themselves provide many different kinds of nourishment for a whole series of fish species. Pointed-jawed butterfly fish get their food parasitically from corals and other stinging animals. They search continuously in the coral stems for small prey caught in the stinging tentacles of coral polyps. As soon as they see these, they produce, by fanning with their pectoral fins, a current so directly aimed at the prey that at the required point a "parting" is made between the polyps, pressing their tentacles flat on all sides and thus enabling the fish to seize the prey almost without getting its nose stung. It always gets it just a little stung and can be seen "sneezing" and shaking its nose, but, like pepper, the sting seems to act as an agreeable stimulant. My beautiful yellow and brown butterfly fishes prefer a prey, such as a peice of fish, stuck in the tentacles of a stinging sea anemone, to the same prey swimming free in the water. Other related species have developed a stronger immunity to stings and they devour the prey together with the coral animal that has caught it. Yet other species disregard the stinging capsules of coelenterates altogether, and eat coral animals, hydroid polyps, and even big, strong, stinging sea anemones, as placidly as a cow eats grass. As well as this immunity to poison, parrot fish have evolved a strong chisellike dentition and they eat whole branches of coral including their calcareous skeleton. If you dive near a grazing herd of these beautiful, rainbow-colored fish, you can hear a cracking and crunching as though a little gravel mill were at work--and this actually corresponds with the facts, for when such a fish excretes, it rains a little shower of white sand, and the observer realizes with astonishment that most of the snow-clean coral sand covering the glades of the coral forest has obviously passed through parrot fish.

Other fish, plectognaths, to which the comical puffers, trunk, and porcupine fish belong, have specialized in cracking hard-shelled mollusks, crabs, and sea urchins; and others again, such as angelfish, specialize in snatching the lovely feather crowns that certain feather worms thrust out of their hard, calcareous tubes. Their capacity for quick retraction acts as a protection against slower predators, but some angelfish have a way of sidling up and, with a lightning sideways jerk of the mouth, seizing the worm's head at a speed surpassing its capacity for withdrawal. Even in the aquarium, where they seize prey which has no such quick reactions, these fish cannot do otherwise than snap like this.

The reef offers many other "openings" for specialized fish. There are some which remove parasites from others and which are therefore left unharmed by the fiercest predators, even when they penetrate right into the mouth cavities of their hosts to perform their hygenic work. There are others which live as parasites on large fish, punching pieces from their epidermis, and among these are the oddest fish of all: they resemble the cleaner fish so closely in color, form, and movement that, under false pretenses, they can safely approach their victims.

It is essential to consider the fact that all these opportunities for special careers, known as ecological niches, are often provided by the same cubic yard of ocean water. Because of the enormous nutritional possibilities, every fish, whatever its specialty, requires only a few square yards of sea bottom for its support, so in this small area there can be as many fish as there are ecological niches, and anyone who has watched with amazement the throngring traffic on a coral reef knows that these are legion. However, every one of this crowd is determined that no other fish of his species should settle in his territory. Specialists of other "professions" harm his livelihood as little as, to use our analogy again, the practice of a doctor harms the trade of a mechanic living in the same village.

In less densely populated biotopes where the same unit of space can support three or four species only, a resident fish or bird can "afford" to drive away all living beings, even members of species that are no real threat to his existence; but if a sedentary coral fish tried to do the same thing, it would be utterly exhausted and, moreover, would never manage to keep its territory free from the swarms of noncompetitors of different "professions." It is in the occupational interests of all sedentary species that each should determine the spatial distribution that will benefit its own individuals, entirely without consideration from other species. The colorful "poster" patterns, described in Chapter One, and the fighting reactions elicited by them, have the effect that the fish of each species keep a measured distance only from nutritional competitors of the same species. This is the very simple answer to the much discussed question of the function of the colors of coral fish.

As I have already mentioned, the species-typical song of birds has a very similar survival value to that of the visual signals of fishes. From the song of a certain bird, other birds not yet in possession of a territory recognize that in this particular place a male is proclaiming territorial rights. It is remarkable that in many species the song indicates how strong and possibly how old the singer is, in other words, how much the listener has to fear him. Among several species of birds that mark their territory acoustically, there is great individual difference of sound expression, and some observers are of the opinion that, in such species, the personal visiting card is of special significance. While Heinroth interpreted the crowing of the cock with the words, "Here is a cock!" Baeumer, the most knowledgeable of all domestic-fowl experts, heard in it the far more special announcement, "Here is the cock Balthazar!"

Among mammals, which mostly "think through their noses," it is not surprising that marking of the territory by scent plays a big role. Many methods have been tried; various scent glands have been evolved, and the most remarkable ceremonies developed around the depositing of urine and feces; of these the leg-lifting of the domestic dog is the most familiar. The objection has been raised by some students of mammals that such scent marks cannot have anything to do with territorial ownership because they are found not only in socially living mammals which do not defend single territories, but also in animals that wander far and wide; but this opinion is only partly correct. First, it has been proved that dogs and other pack-living animals recognize each other by the scent of the marks, and it would at once be apparent to the members of a pack if a nonmember presumed to lift its leg in their hunting grounds. Secondly, Leyhausen and Wolf have demonstrated the very interesting possibility that the distribution of animals of a certain species over the available biotope can be effected not only by a space plan but also by a time plan. They found that, in domestic cats living free in open country, several individuals could make use of the same hunting ground without even coming into conflict, by using it according to a definite time-table, in the same way as our Seewiesen housewives use our communal washhouse. An additional safeguard against undesirable encounters is the scent marks which these animals--the cats, not the housewives--deposit at regular intervals wherever they go. These act like railway signals whose aim is to prevent collision between two trains. A cat finding another cat's signal on its hunting path assesses its age, and if it is very fresh it hesitates, or chooses another path; if it is a few hours old it proceeds calmly on its way.

Even in the case of animals whose territory is governed by space only, the hunting ground must not be imagined as a property determined by geographical confines; it is determined by the fact that in every individual the readiness to fight is least diminished by its readiness to escape. As the distance from this "headquarters" increases, the readiness to fight decreases proportionately as the surroundings become stranger and more intimidating to the animal. If one plotted the graph of this decrease the curve would not be equally steep for all directions in space. In fish, the center of whose territory is nearly always on the bottom, the decline in readiness to fight is most marked in the vertical direction because the fish is threatened by special dangers from above.

The territory which an animal apparently possesses is thus only a matter of variations in readiness to fight, depending on the place and on various local factors inhibiting the fighting urge. In nearing the center of the territory the aggressive urge increases in geometrical ratio to the decrease in distance from this center. This increase in aggression is so great that it compensates for all differences ever to be found in adult, sexually mature animals of a species. If we know the territorial centers of two conflicting animals, such as two garden redstarts or two aquarium sticklebacks, all other things being equal, we can predict, from the place of encounter, which one will win: the one that is nearer home.

When the loser flees, the inertia of reaction of both animals leads to that phenomenon which always occurs when a time lag enters into a self-regulating process-- to an oscillation. The courage of the fugitive returns as he nears his own headquarters, while that of the pursuer sinks in proportion to the distance covered in enemy territory. Finally the fugitive turns and attacks the former pursuer vigorously and unexpectedly and, as was predictable, he in his turn is beaten and driven away. The whole performance is repeated several times till both fighters come to a standstill at a certain point of balance where they threaten each other without fighting.

The position, the territorial "border," is in no way marked on the ground but is determined exclusively by a balance of power and may, if this alters in the least, for instance if one fish is replete and lazy, come to lie in a new position somewhat nearer the headquarters of the lazy one. An old record of our observations on the territorial behavior of two pairs of cichlids demonstrates this oscillation of the territorial borders. Four fish of this species were put into a large tank and at once the strongest male, A, occupied the left, back, lower corner and chased the other three mercilessly around the whole tank; in other words, he claimed the whole tank as his territory. After a few days, male B took possession of a tiny space immediately below the surface in the diagonally opposite right, front, upper corner. There he bravely resisted the attacks of the first male. This occupation of an area near the surface is in a way an act of desperation for one of these fish, because it is risking great danger from aerial predators in order to hold its own against an enemy of its own species, which, as already explained, will attack less resolutely in such a locality. In other words, the owner of such a dangerous area has, as an ally, the fear which the surface inspires in its bad neighbor. During succeeding days, the space defended by B grew visibly, expanding downward until he finally took his station in the right, front, lower corner, so gaining a much more satisfactory headquarters. Now at last he had the same chances as A, whom he quickly pressed so far back that their territories divided the tank into two almost equal parts. It was interesting to see how both fishes patrolled the border continuously, maintaining a threatening attitude. Then one morning they were doing this on the extreme right of the tank, again around the original headquarters of B, who could now scarcely call a few square inches his own. I knew at once what had happened: A had paired, and since it is characteristic of all large cichlids that both partners take part in territorial defense, B was subjected to double pressure and his territory had decreased accordingly. Next day the fish were again in the middle of the tank, threatening each other across the "border," but now there were four, because B had also taken a mate, and thus the balance of power with the A family was restored. A week later I found the border far toward the left lower area, and encroaching on A's former territory. The reason for this was that the A couple had spawned and since one of the partners was busy looking after the eggs, only one at a time was able to attend to frontier defense. As soon as the B couple had also spawned, the previous equal division of space was re-established. Julian Huxley once used a good metaphor to describe this behavior: he compared the territories to air-balloons in a closed container, pressing against each other and expanding or contracting with the slightest change of pressure in each individual one. This territorial aggression, really a very simple mechanism of behavior-physiology, gives an ideal solution to the problem of the distribution of animals of any one species over the available area in such a way that it is favorable to the species as a whole. Even the weaker specimens can exist and reproduce, if only in a very small space. This has special significance in creatures which reach sexual maturity long before they are fully grown. What a peaceful issue of the "evil principle"!

In many animals the same result is achieved without aggressive behavior. Theoretically it suffices that animals of the same species "cannot bear the smell of each other" and avoid each other accordingly. To a certain extent this applies to the smell signals deposited by cats, though behind these lies a hidden threat of active aggression. There are some vertebrates which entirely lack intra-specific aggression but which nevertheless avoid their own species meticulously. Some frogs, in particular tree frogs, live solitary lives except at mating time, and they are obviously distributed very evenly over the available habitat. As American scientists have recently discovered, this distribution is effected quite simply by the fact that every frog avoids the quacking sound of his own species. This explanation, however, does not account for the distribution of the females, for these, in most frogs, are dumb.

We can safely assume that the most important function of intra-specific aggression is the even distribution of the animals of a particular species over an inhabitable area, but it is certainly not its only one. Charles Darwin had already observed that sexual selection, the selection of the best and strongest animals for reproduction, was furthered by the fighting of rival animals, particularly males. The strength of the father directly affects the welfare of the children in those species in which he plays an active part in their care and defense. The correlation between male parental care and rival fighting is clear, particularly in those animals which are not territorial in the sense which the Cichlids demonstrate but which wander more or less nomadically, as, for example, large ungulates, ground apes, and many others. In such animals, intra-specific aggression plays no essential part in the "spacing out" of the species. Bisons, antelopes, horses, etc., form large herds, and territorial borders and territorial jealousy are unknown to them since there is enough food for all. Nevertheless the males of these species fight each other violently and dramatically, and there is no doubt that the selection resulting from this aggressive behavior leads to the evolution of particularly strong and courageous defenders of family and herd; conversely, there is just as little doubt that the survival value of herd defense has resulted in selective breeding for hard rival fights. This interaction has produced impressive fighters such as bull bison or the males of the large baboon species; at every threat to the community, these valiantly surround and protect the weaker members of the herd.

In connection with rival fights attention must be drawn to a fact which, though it seems paradoxical to the nonbiologist, is, as we shall show later on in this book, of the very greatest importance: purely intra-specific selective breeding can lead to the development of forms and behavior patterns which are not only nonadaptive but can even have adverse effects on species preservation. This is why, in the last paragraph, I emphasized the fact that family defense, a form of strife with the extra-specific environment, has evolved the rival fight, and this in its turn has developed the powerful males. If sexual rivalry, or any other form of intra-specific competition, exerts selection pressure uninfluenced by any environmental exigencies, it may develop in a direction which is quite unadaptive to environment, and irrelevant, if not positively detrimental, to survival. This process may give rise to bizarre physical forms of no use to the species. The antlers of stags, for example, were developed in the service of rival fights, and a stag without them has little hope of producing progeny. Otherwise antlers are useless, for male stags defend themselves against beasts of prey with their fore-hoofs only and never with their antlers. Only the reindeer has based an invention on this necessity and "learned" to shovel snow with a widened point of its antlers.

Sexual selection by the female often has the same results as the rival fights. Wherever we find exaggerated development of colorful feathers, bizarre forms, etc., in the male, we may suspect that the males no longer fight but that the last word in the choice of a mate is spoken by the female, and that the male has no means of contesting this decision. Birds of Paradise, the Ruff, the Mandarin Duck, and the Argus Pheasant show examples of such behavior. The Argus hen pheasant reacts to the large secondary wing feathers of the cock; they are decorated with beautiful eye spots and the cock spreads them before her during courtship. They are so huge that the cock can scarcely fly, and the bigger they are the more they stimulate the hen. The number of progeny produced by a cock in a certain period of time is in direct proportion to the length of these feathers, and, even if their extreme development is unfavorable in other ways--his unwieldiness may cause him to be eaten by a predator while a rival with less absurdly exaggerated wings may escape--he will nevertheless leave more descendants than will a plainer cock. So the predisposition to huge wing feathers is preserved, quite against the interests of the species. One could well imagine an Argus hen that reacted to a small red spot on the wings of the male, which would disappear when he folded his wings and interfere neither with his flying capacity nor with his protective color, but the evolution of the Argus pheasant has run itself into a blind alley. The males continue to compete in producing the largest possible wing feathers, and these birds will never reach a sensible solution and "decide" to stop this nonsense at once.

Here for the first time we are up against a strange and almost uncanny phenomenon. We know that the techniques of trial and error used by the great master builders sometimes lead inevitably to plans that fall short of perfect efficiency. In the plant and animal worlds there are, besides the efficient, quantities of characteristics which only just avoid leading the particular species to destruction. But in the case of the Argus pheasant we have something quite different: it is not only like the strict efficiency expert "closing an eye" and letting second-rate construction pass in the interests of experiment, but it is selection itself that has here run into a blind alley which may easily result in destruction. This always happens when competition between members of a species causes selective breeding without any relation to the extra-specific environment.

My teacher, Oskar Heinroth, used to say jokingly, "Next to the wings of the Argus pheasant, the hectic life of Western civilized man is the most stupid product of intra-specific selection!" The rushed existence into which industrialized, commercialized man has precipitated himself is actually a good example of an inexpedient development caused entirely by competition between members of the same species. Human beings of today are attached by so-called manager diseases, high blood pressure, renal atrophy, gastric ulcers, and torturing neuroses; they succumb to barbarism because they have no more time for cultural interests. And all this is unnecessary, for they could easily agree to take things more easily; theoretically they could, but in practice it is just as impossible for them as it is for the Argus pheasant to grow shorter wing feathers.

There are still worse consequences of intra-specific selection, and for obvious reasons man is particularly exposed to them: unlike any creature before him, he has mastered all hostile powers in his environment, he has exterminated the bear and the wolf and now, as the Latin proverb says, "Homo homini lupus." Striking support for this view comes from the work of modern American sociologists, and in his book The Hidden Persuaders Vance Packard gives an impressive picture of the grotesque state of affairs to which commercial competition can lead. Reading this book, one is tempted to believe that intra-specific competition is the "root of all evil: in a more direct sense than aggression can ever be.

In this chapter on the survival value of aggression, I have laid special stress on the potentially destructive effects of intra-specific selection: because of them, aggressive behavior can, more than other qualities and functions, become exaggerated to the point of the grotesque and inexpedient. In later chapters we shall see what effects it has had in several animals, for example, in the Egyptian Goose and the Brown Rat. Above all, it is more than probable that the destructive intensity of the aggression drive, still a hereditary evil of mankind, is the consequence of a process of intra-specific selection which worked on our forefathers for roughly forty thousand years, that is, throughout the Early Stone Age. When man had reached the stage of having weapons, clothing, and social organization, so overcoming the dangers of starving, freezing, and being eaten by wild animals, and these dangers ceased to be the essential factors influencing selection, an evil intra-specific selection must have set in. The factor influencing selection was now the wars waged between hostile neighboring tribes. These must have evolved in an extreme form of all those so-called "warrior virtues" which unfortunately many people still regard as desirable ideals. We shall come back to this in the last chapter of this book.

I return to the theme of the survival value of the rival fight, with the statement that this only leads to useful selection where it breeds fighters fitted for combat with extra-specific enemies as well as for intra-specific duels. The most important function of rival fighting is the selection of an aggressive family defender, and this presupposes a further function of intra-specific aggression: brood defense. This is so obvious that it requires no further comment. If it should be doubted, its truth can be demonstrated by the fact that in many animals, where only one sex cares for the brood, only that sex is really aggressive toward fellow members of the species. Among sticklebacks it is the male, in several dwarf cichlids the female. In many gallinaceous birds, only the female tend the brood, and these are often far more aggressive than the males. The same thing is said to be true of human beings.

It would be wrong to believe that the three functions of aggressive behavior dealt with in the last three chapters--namely, balanced distribution of animals of the same species over the available environment, selection of the strongest by rival fights, and defense of the young--are its only important functions in the preservation of the species. We shall see later what an indispensable part in the great complex of drives is played by aggression; it is one of those driving powers which students of behavior call "motivation"; it lies behind behavior patterns that outwardly have nothing to do with aggression, and even appear to be its very opposite. It is hard to say whether it is a paradox or a commonplace that, in the most intimate bonds between living creatures, there is a certain measure of aggression. Much more remains to be said before discussing this central problem in our natural history of aggression. The important part played by aggression in the interaction of drives within the organism is not easy to understand and still less easy to expound.

We can, however, here describe the part played by aggression in the structure of society among highly developed animals. Though many individuals interact in a social system, its inner workings are often easier to understand than the interaction of drives within the individual. A principle of organization without which a more advanced social life cannot develop in higher vertebrates is the so-called ranking order. Under this rule every individual in the society knows which one is stronger and which weaker than itself, so that everyone can retreat from the stronger and expect submission from the weaker, if they should get in each other's way. Schjelderup-Ebbe was the first to examine the ranking order in the domestic fowl and to speak of the "pecking order," an expression used to this day by writers. It seems a little odd though, to me, to speak of a pecking order even for large animals which certainly do not peck, but bite or ram. However, its wide distribution speaks for its great survival value, and therefore we must ask wherein this lies.

The most obvious answer is that it limits fighting between the members of a society, but here in contrast one may ask: Would it not have been better if aggression among members of a society were utterly inhibited? To this, a whole series of answers can be given. First, as we shall discuss very thoroughly in a later chapter (Ten, "The Bond"), the case may arise that a society, for example, a wolf pack or monkey herd, urgently needs aggression against other societies of the same species, therefore aggression should be inhibited only <u>inside</u> the horde. Secondly, a society may derive a beneficial firmness of structure from the state of tension arising inside the community from the aggression drive and its result, ranking order. In jackdaws, and in many other very social birds, ranking order leads directly to protection of the weaker ones. All social animals are "status seekers," hence there is always particularly high tension between individuals who held immediately adjoining positions in the ranking order; conversely, this tension diminishes the further apart the two animals are in rank. Since high-ranking jackdaws, particularly males, interfere in every quarrel between two inferiors, this graduation of social tension has the desirable effect that the higher-ranking birds always intervene in favor of the losing party.

In jackdaws, another form of "authority" is already linked with the ranking position which the individual has acquired by its aggressive drive. The expression movements of a high-ranking jackdaw, particularly of an old male, are given much more attention by the colony members than those of a lower-ranking, young bird. For example, if a young bird shows fright at some meaningless stimulus, the others, especially the older ones, pay almost no attention to his expressions of fear. But if the same sort of alarm proceeds from one of the old males, all the jackdaws within sight and earshot immediately take flight. Since, in jackdaws, recognition of predatory enemies is not innate but is learned by every individual from the behavior of experienced old birds, it is probably of considerable importance that great store is set by the "opinion" of old, high-ranking, and experienced birds.

With the higher evolution of an animal species, the significance of the role played by individual experience and learning generally increases, while innate behavior, though not losing importance, becomes reduced to simpler though not less numerous elements. With this general trend in evolution, the significance attached to the experienced old animal becomes greater all the time, and it may even be said that the social coexistence of intelligent mammals has achieved a new survival value by the use it makes of the handing down of individually acquired information. Conversely, it may be said that social coexistence exerts selection pressure in the direction of better learning capacity, because in social animals this faculty benefits not only the individual but also the community. Thus longevity far beyond the age of reproductive capacity has considerable species-preserving value. We know from Fraser Darling and Margaret Altmann that in many species of deer the herd is led by an aged female, no longer hampered in her social duties by the obligations of motherhood.

All other conditions being equal, the age of an animal is, very consistently, in direct proportion to the position it holds in the ranking order of its society. It is thus advantageous if the "constructors" of behavior rely upon this consistency and if the members of the community--who cannot read the age of the experienced leader animal in its birth certificate--rate its reliability by its rank. Some time ago, collaborators of Robert M. Yerkes made the extraordinarily interesting observation that chimpanzees, animals well known to be capable of learning by imitation, copy only higher-ranking members of their species. From a group of these apes, a low-ranking individual was taken and taught to remove bananas from a specially constructed feeding apparatus by very complicated manipulations. When this ape, together with his feeding apparatus, was brought back to the group, the higher-ranking animals tried to take away the bananas which he had acquired for himself, but none of them thought of watching their inferior at work and learning something from him. Then the highest-ranking chimpanzee was removed and taught to use the apparatus in the same way, and when he was put back in the group the other members watched him with great interest and soon learned to imitate him.

S. L. Washburn and Irven de Vore observed that among free-living baboons the band was led not by a single animal but by a "senate" of several old males who maintained their superiority over the younger and physically stronger members by firmly sticking together and proving, as a united force, stronger than any single young male. In a more exactly observed case, one of the three "senators" was seen to be an almost toothless old creature while the other two were well past their prime. On one occasion when the band was in a treeless area and in danger of encountering a lion, the animals stopped and the young, strong males formed a defensive circle around the weaker animals. But the oldest male went forward alone, performed the dangerous task of finding out exactly where the lion was lying, without being seen by him, and then returned to the horde and led them, by a wide detour around the lion, to the safety of their sleeping trees. All followed him blindly, no one doubting his authority.

Let us look back on all that we have learned in this chapter from the objective observation of animals, and consider in what ways intra-specific aggression assists the preservation of an animal species. The environment is divided between the members of the species in such a way that, within the potentialities offered, everyone can exist. The best father, the best mother are chosen for the benefit of the progeny. The children are protected. The community is so organized that a few wise males, the "senate," acquire the authority essential for making and carrying out decisions for the good of the community. Though occasionally, in territorial or rival fights, by some mishap a horn may penetrate an eye or a tooth an artery, we have never found that the aim of aggression was the extermination of fellow members of the species concerned. This of course does not negate the fact that under unnatural circumstances, for example confinement, unforeseen by the "constructors" of evolution, aggressive behavior may have a destructive effect.

Let us now examine ourselves and try, without selfconceit but also without regarding ourselves as miserable sinners, to find out what we would like to do, in a state of highest violent aggressive feeling, to the person who elicited that emotion. I do not think I am claiming to be better than I am when I say that the final, drive-assuaging act, Wallace Craig's consummatory act, is not the killing of my enemy. The satisfying experience consists, in such cases, in administering a good beating, but certainly not in shooting or disemboweling; and the desired objective is not that my opponent should lie dead but that he should be soundly thrashed and humbly accept my physical and, if I am to be considered as good as a baboon, my mental superiority. And since, on principle, I only wish to thrash such fellows as deserve these humiliations, I cannot entirely condemn my instincts in this connection. However, it must be admitted that a slight deviation from nature, a coincidence that put a knife into one's hand at the critical moment, might turn as intended thrashing into manslaughter.

Summing up what has been said in this chapter, we find that aggression, far from being the diabolical, destructive principle that classical psychoanalysis makes it out to be, is really an essential part of the life-preserving organization of instincts. Though by accident it may function in the wrong way and cause destruction, the same is true of practically any functional part of any system. Moreover, we have not yet considered an all-important fact which we shall hear about in Chapter Ten. Mutation and selection, the great "constructors" which make genealogical trees grow upward, have chosen, of all unlikely things, the rough and spiny shoot of intra-specific aggression to bear the blossoms of personal friendship and love.

Two Poems
William Blake

    Lorenz's closing words on the evolutionary roots of love brought two short poems to mind. Both of them were written by William Blake.

The Angel
that presided
o'er my
birth said
"Little creature,
form'd of
Joy and Mirth,
Go love
without the
help of
any thing
on Earth."

Love seeketh only Self to please,
To bind another to Its delight,
Joys in another's loss of ease,
And Builds a Hell in Heaven's despite.

*

THE END OF THE SPECIES
Pierre Teilhard de Chardin

Pierre Teilhard de Chardin was a paleontologist, a priest, and a brilliant philosopher. His spiritual vision of evolving man is as exciting as it is comforting. If you have trouble with his terminology, stay with him anyway. His message is well worth the effort.

Not much more than a hundred years ago Man learned to his astonishment that there was an origin of animal species, a genesis in which he himself was involved. Not only did all kinds of animals share the earth with him, but he found that he was in some sort a part of this zoological diversity which hitherto he had regarded as being merely his neighbours. Life was in movement, and Mankind was the latest of its successive waves!

This astonishing pronouncement on the part of science seemed at first to do no more than stimulate the curiosity (or indignation) of theorists; but it was soon apparent that the shock was not purely mental, and that nineteenth century man had been shaken by it to his depths. Three hundred years earlier, in the time of Galileo, the end of geocentrism had intrigued or disturbed thinking minds without having any appreciable effect on the mass of people. The sidereal dispute had, after all, produced no change in the earth itself, or in its inhabitants or their relations with one another. But the concept of biological evolution inevitably led to a profound reshaping of planetary values.

To some outraged spirits, no doubt, Man appeared diminished and dethroned by this evolutionary theory which made him no more than the latest arrival in the

---

"The End of the Species" in *The Future of Man* by Pierre Teilhard de Chardin; translated by Norman Denny. Copyright 1959 by Editions du Deuil. Copyright © 1964 in the English translation by Wm. Collins Sons & Co. Ltd., London and Harper & Row, Publishers, Inc., New York. Reprinted by permission of Harper & Row, Publishers, Inc.

animal kingdom. But to the minds of the majority our human condition seemed finally to be exalted by the fact that we were rooted in the fauna and soil of the planet-- evolving Man in the forefront of the animals.

In short, until then Man, although he knew that the human race might continue to exist for a long time, had not suspected that it had a future. Now however, because he was a species, and species change, he could begin to look for and seek to conquer something quite new that lay ahead of him.

That is why 'Darwinism', as it was then called, however naive its beginnings, came at exactly the right moment to create the cosmological atmosphere of which the great technicosocial advance of the last century stood in need if it was to believe passionately in what it was doing. Rudimentary though it was, Darwinism afforded a scientific justification of faith in progress.

But today, by a development natural to itself, the movement has come to look like a receding tide. For all his discoveries and inventions, twentieth century man is a sad creature. How shall we account for his present dejected state except basically by the fact that, following that exalted vision of species in growth, he is now confronted by an accumulation of scientific evidence pointing to the reverse--the species doomed to extinction?

The extinction of the species...

Biologists do not agree about the mechanism of the continual disappearance of phyla in the course of geological time, a process almost as mysterious as that of their formation; but the reality of the phenomenon is indisputable. Either the different species, losing their powers of 'speciation', survive as living fossils, which after all is a form of death; or else, and there are infinitely more of these, they simply vanish, one sort being replaced by another. Whatever the reason may be, inadaptability to a new environment, competition, a mysterious senescence, or possibly a single basic cause underlying all these reasons, the end is always the same. The days (or the millennia) of every living form are by statistical reckoning ineluctably numbered; so much so that, using the scale of time furnished by the study of certain isotopes, it is beginning to be possible to calculate in millions of years the average life of a species.

Man now sees that the seeds of his ultimate dissolution are at the heart of his being. The End of the Species is in the marrow of our bones!

Is it not this presentiment of a blank wall ahead, underlying all other tensions and specific fears, which paradoxically (at the very moment when every barrier seems to be giving way before our power of understanding and mastering the world) is darkening and hardening the minds of our generation?

As psychiatry teaches us, we shall gain nothing by shutting our eyes to this shadow of collective death that has appeared on our horizon. On the contrary, we must open them wider.

But how are we to exorcise the shadow?

It may be said that timidly, even furtively (it is remarkable how coy we are in referring to the matter) two methods are used by writers and teachers to reassure themselves and others in face of the ever more obsessive certainty of the eventual ending of the human species: the first is to invoke the infinity of Time and the second is to seek shelter in the depths of Space.

The Time argument is as follows. By the latest estimates of palaeontology the probable life of a phylum of average dimensions is to be reckoned in tens of millions of years. But if this is true of 'ordinary' species, what duration may we not look for in the case of Man, that favoured race which, by its intelligence, has succeeded in removing all danger of serious competition and even in attacking the causes of senescence at the root.

Then the Space argument. Even if we suppose that, by prolonging its existence on a scale of planetary longevity, the human species will eventually find itself with a chemically exhausted Earth beneath its feet, is not Man even now in process of developing astronautical means which will enable him to go elsewhere and continue his destiny in some other corner of the firmament?

That is what they say, and for all I know there may be people for whom this sort of reasoning does really dispel the clouds that veil the future. I can only say that for my part I find such consolations intolerable, not only because they do nothing but palliate and postpone our fears, which is bad enough, but even more because they seem to me scientifically false.

In order that the end of Mankind may be deferred <u>sine die</u> we are asked to believe in a species that will drag on and spread itself indefinitely; which means, in effect, that it would run down more and more. But is not this the precise opposite of what is happening here and now in the human world?

I have been insisting for a long time on the importance and significance of the technico-mental process which, particularly during the past hundred years, has been irresistibly causing Mankind to draw closer together and unite upon itself. From routine or prejudice the majority of anthropologists still refuse to see in this movement of totalisation anything more than a superficial and temporary side-effect of the organic forces of biogenesis. Any parallel that may be drawn between socialisation and speciation, they maintain, is purely metaphorical. To which I would reply that, if this is so, to what undisclosed form of energy shall we scientifically attribute the irreversible and conjugated growth of Arrangement and Consciousness which historically characterises (as it does everywhere else, in indisputably 'biological' fields) the establishment of Mankind on Earth?

We have only to go a little further, I am convinced, and our minds, awakened at last to the existence of an added dimension, will grasp the profound identity existing between the forces of civilisation and those of evolution. Man will then assume his true shape in the eyes of the naturalists--that of a species which, having entered the realm of Thought, henceforth folds back its branches upon itself instead of spreading them. Man, <u>a species which converges</u>, instead of diverging like every other species on earth: so that we are bound to envisage its ending in terms of some paroxysmal state of maturation which, by its scientific probability alone, must illumine for us all the darkest menaces of the future.

For if by its structure Mankind does not dissipate itself but concentrates upon itself; in other words, if, alone among all the living forms known to us, our zoological phylum is laboriously moving towards a <u>critical point of speciation</u>, then are not all hopes permitted to us in the matter of survival and irreversibility?

The end of a 'thinking species': not disintegration and death, but a new break-through and a re-birth, this time outside Time and Space, through the very excess of unification and coreflexion.[1]

It goes without saying that this idea of a salvation of the Species sought, not in the direction of any temporo-spatial consolidation or expansion but by way of spiritual escape through the excess of consciousness, is not yet seriously considered by the biologists. At first sight it appears fantastic. Yet if one thinks about it

---

[1] Such reflexion, as I am constantly obliged to say, in no way entailing a diminution but on the contrary an increase of the 'person'. Must I again repeat the truth, of universal application, that if it be properly ordered union <u>does not confound but differentiates</u>?

long and carefully, it is remarkable how it sustains examination, grows stronger and, for two particular reasons among others, takes root in the mind.

For one thing, as I have said, it corresponds more closely than any other extrapolation to the marked (even challenging) urgency of our own time in the broad progress of the Phenomenon of Man. But in addition it seems to be more capable than any other vision of the future of stimulating and steadying our power of action by counteracting the prevailing pessimism.

Finally, there is a fact which we must face.

In the present age, what does most discredit to faith in progress (apart from our reticence and helplessness as we contemplate the 'end of the Race') is the unhappy tendency still prevailing among its adepts to distort everything that is most valid and noble in our newly aroused expectation of an 'ultra-human' by reducing it to some form of threadbare millennium. The believers in progress think in terms of a Golden Age, a period of euphoria and abundance; and this, they give us to understand, is all that Evolution has in store for us. It is right that our hearts should fail us at the thought of so 'bourgeois' a paradise.

We need to remind ourselves yet again, so as to offset this truly pagan materialism and naturalism, that although the laws of biogenesis by their nature presuppose, and in fact bring about, an improvement in human living conditions, it is not <u>well-being</u> but a hunger for <u>more-being</u> which, of psychological necessity, can alone preserve the thinking earth from the <u>taedium vitae</u>. And this makes fully plain the importance of what I have already suggested, that it is upon its point (or superstructure) of spiritual concentration, and not on its basis (or infrastructure) of material arrangement, that the equilibrium of Mankind biologically depends.

For if, pursuing this thought, we accept the existence of a critical point of speciation at the conclusion of all technologies and civilisations, it means (with Tension maintaining its ascendancy over Rest to the end of biogenesis) that an <u>outlet</u> appears at the peak of Time, not only for our hope of escape but for our expectation of revelation.

And this is what can best allay the conflict between light and darkness, exaltation and despair, in which, following the rebirth in us of the Sense of Species, we are now absorbed.

PART III
ECOLOGY

THE HISTORICAL ROOTS OF OUR ECOLOGIC CRISIS
Lynn White, Jr.

Lynn White finds the basis of our environmental dilema in an alienation from nature deeply grounded in Christian dogma.

A conversation with Aldous Huxley not infrequently put one at the receiving end of an unforgettable monologue. About a year before his lamented death he was discoursing on a favorite topic: Man's unnatural treatment of nature and its sad results. To illustrate his point he told how, during the previous summer, he had returned to a little valley in England where he had spent many happy months as a child. Once it had been composed of delightful grassy glades; now it was becoming overgrown with unsightly brush because the rabbits that formerly kept such growth under control had largely succumbed to a disease, myxomatosis, that was deliberately introduced by the local farmers to reduce the rabbits' destruction of crops. Being something of a Philistine, I could be silent no longer, even in the interests of great rhetoric. I interrupted to point out that the rabbit itself had been brought as a domestic animal to England in 1176, presumably to improve the protein diet of the peasantry.

All forms of life modify their contexts. The most spectacular and benign instance is doubtless the coral polyp. By serving its own ends, it has created a vast undersea world favorable to thousands of other kinds of animals and plants. Ever since man became a numerous species he has affected his environment notably. The hypothesis that his fire-drive method of hunting created the world's great grasslands and helped to exterminate the monster mammals of the Pleistocene from much of the globe is plausible, if not proved. For 6 millennia at least, the banks of the

Reprinted by permission from *Science*, Vol. 155, pp. 1203-1207, March 10, 1968, copyright 1968 by the American Association for the Advancement of Science.

lower Nile have been a human artifact rather than the swampy African jungle which nature, apart from man, would have made it. The Aswan Dam, flooding 5000 square miles, is only the latest stage in a long process. In many regions terracing or irrigation, overgrazing, the cutting of forests by Romans to build ships to fight Carthaginians or by Crusaders to solve the logistics problems of their expeditions, have profoundly changed some ecologies. Observation that the French landscape falls into two basic types, the open fields of the north and the bocage of the south and west, inspired Marc Bloch to undertake his classic study of medieval agricultural methods. Quite unintentionally, changes in human ways often affect nonhuman nature. It has been noted, for example, that the advent of the automobile eliminated huge flocks of sparrows that once fed on the horse manure littering every street.

The history of ecologic change is still so rudimentary that we know little about what really happened, or what the results were. The extinction of the European aurochs as late as 1627 would seem to have been a simple case of overenthusiastic hunting. On more intricate matters it often is impossible to find solid information. For a thousand years or more the Frisians and Hollanders have been pushing back the North Sea, and the process is culminating in our own time in the reclamation of the Zuider Zee. What, if any, species of animals, birds, fish, shore life, or plants have died out in the process? In their epic combat with Neptune have the Netherlanders overlooked ecological values in such a way that the quality of human life in the Netherlands has suffered? I cannot discover that the questions have ever been asked, much less answered.

People, then, have often been a dynamic element in their own environment, but in the present state of historical scholarship we usually do not know exactly when, where, or with what effects man-induced changes came. As we enter the last third of the 20th century, however, concern for the problem of ecologic backlash is mounting feverishly. Natural science, conceived as the effort to understand the nature of things, had flourished in several eras and among several peoples. Similarly there had been an age-old accumulation of technological skills, sometimes growing rapidly, sometimes slowly. But it was not until about four generations ago that Western Europe and North America arranged a marriage between science and technology, a union of the theoretical and the empirical approaches to our natural environment. The emergence in widespread practice of the Baconian creed that scientific knowledge means technological power over nature can scarcely be dated before about 1850, save in the chemical industries, where it is anticipated in the 18th century. Its acceptance as a normal pattern of action may mark the greatest event in human history since the invention of agriculture, and perhaps in nonhuman terrestrial history as well.

Almost at once the new situation forced the crystallization of the novel concept of ecology; indeed, the word ecology first appeared in the English language in 1873. Today, less than a century later, the impact of our race upon the environment has so increased in force that it has changed in essence. When the first cannons were fired, in the early 14th century, they affected ecology by sending workers scrambling to the forests and mountains for more potash, sulfur, iron ore, and charcoal, with some resulting erosion and deforestation. Hydrogen bombs are of a different order: a war fought with them might alter the genetics of all life on this planet. By 1285 London had a smog problem arising from the burning of soft coal, but our present combustion of fossil fuels threatens to change the chemistry of the globe's atmosphere as a whole, with consequences which we are only beginning to guess. With the population explosion, the carcinoma of planless urbanism, the now geological deposits of sewage and garbage, surely no creature other than man has ever managed to foul its nest in such short order.

There are many calls to action, but specific proposals, however worthy as individual items, seem too partial, palliative, negative: ban the bomb, tear down the billboards, give the Hindus contraceptives, and tell them to eat their sacred cows. The simplest solution to any suspect change is, of course, to stop it, or, better yet, to revert to a romanticized past: make those ugly gasoline stations look like Anne Hathaway's cottage or (in the Far West) like ghost-town saloons. The "wilderness area" mentality invariably advocates deep-freezing an ecology, whether San Gimignano or the High Sierra, as it was before the first Kleenex was dropped. But neither atavism nor prettification will cope with the ecologic crisis of our time.

What shall we do? No one yet knows. Unless we think about fundamentals, our specific measures may produce new backlashes more serious than those they are designed to remedy.

As a beginning we should try to clarify our thinking by looking, in some historical depth, at the presuppositions that underlie modern technology and science. Science was traditionally aristocratic, speculative, intellectual in intent; technology was lower-class, empirical, action-oriented. The quite sudden fusion of these two, toward the middle of the 19th century, is surely related to the slightly prior and contemporary democratic revolutions which, by reducing social barriers, tended to assert a functional unity of brain and hand. Our ecologic crisis is the product of an emerging, entirely novel, democratic culture. The issue is whether a democratized world can survive its own implications. Presumably we cannot unless we rethink our axioms.

THE WESTERN TRADITIONS OF TECHNOLOGY AND SCIENCE

One thing is so certain that it seems stupid to verbalize it: both modern technology and modern science are distinctively Occidental. Our technology has absorbed elements from all over the world, notably from China; yet everywhere today, whether in Japan or in Nigeria, successful technology is Western. Our science is the heir to all the sciences of the past, especially perhaps to the work of the great Islamic scientists of the Middle Ages, who so often outdid the ancient Greeks in skill and perspicacity: al-Razi in medicine, for example; or ibn-al-Haytham in optics; or Omar Khayyam in mathematics. Indeed, not a few works of such geniuses seem to have vanished in the original Arabic and to survive only in medieval Latin translations that helped to lay the foundations for later Western developments. Today, around the globe, all significant science is Western in style and method, whatever the pigmentation or language of the scientists.

A second pair of facts is less well recognized because they result from quite recent historical scholarship. The leadership of the West, both in technology and in science, is far older than the so-called Scientific Revolution of the 17th century or the so-called Industrial Revolution of the 18th century. These terms are in fact outmoded and obscure the true nature of what they try to describe--significant stages in two long and separate developments. By A.D. 1000 at the latest--and perhaps, feebly, as much as 200 years earlier--the West began to apply water power to industrial processes other than milling grain. This was followed in the late 12th century by the harnessing of wind power. From simple beginnings, but with remarkable consistency of style, the West rapidly expanded its skills in the development of power machinery, labor-saving devices, and automation. Those who doubt should contemplate that most monumental achievement in the history of automation: the weight-driven mechanical clock, which appeared in two forms in the early 14th century. Not in craftsmanship but in basic technological capacity, the Latin West of the later Middle Ages far outstripped its elaborate, sophisticated, and esthetically magnificent sister cultures, Byzantium and Islam. In 1444 a great Greek ecclesiastic, Bessarion, who had gone to Italy, wrote a letter to a prince in Greece. He is amazed by the superiority of Western ships, arms, textiles, glass. But above all he is astonished by the spectacle of water-wheels sawing timbers and pumping the bellows of blast furnaces. Clearly, he had seen nothing of the sort in the Near East.

By the end of the 15th century the technological superiority of Europe was such that its small, mutually hostile nations could spill out over all the rest of the world, conquering, looting, and colonizing. The symbol of this technological superiority is the fact that Portugal, one of the weakest states of the Occident, was able to become, and to remain for a century, mistress of the East Indies. And we must remember that the technology of Vasco da Gama and Albuquerque was built by pure empiricism, drawing remarkably little support or inspiration from science.

In the present-day vernacular understanding, modern science is supposed to have begun in 1543, when both Copernicus and Vesalius published their great works. It is no derogation of their accomplishments, however, to point out that such structures as the Fabrica and the De revolutionibus do not appear overnight. The distinctive Western tradition of science, in fact, began in the late 11th century with a massive movement of translation of Arabic and Greek scientific works into Latin. A few notable books--Theophrastus, for example--escaped the West's avid new appetite for science, but within less than 200 years effectively the entire corpus of Greek and Muslim science was available in Latin, and was being eagerly read and criticized in the new European universities. Out of criticism arose new observation, speculation, and increasing distrust of ancient authorities. By the late 13th century Europe had seized global scientific leadership from the faltering hands of Islam. It would be as absurd to deny the profound originality of Newton, Galileo, or Copernicus as to deny that of the 14th century scholastic scientists like Buridan or Oresme on whose work they built. Before the 11th century, science scarcely existed in the Latin West, even in Roman times. From the 11th century onward, the scientific sector of Occidental culture has increased in a steady crescendo.

Since both our technological and our scientific movements got their start, acquired their character, and achieved world dominance in the Middle Ages, it would seem that we cannot understand their nature or their present impact upon ecology without examining fundamental medieval assumptions and developments.

MEDIEVAL VIEW OF MAN AND NATURE

Until recently, agriculture has been the chief occupation even in "advanced" societies; hence, any change in methods of tillage has much importance. Early plows, drawn by two oxen, did not normally turn the sod but merely scratched it. Thus, cross-plowing was needed and fields tended to be squarish. In the fairly light soils and semiarid climates of the Near East and Mediterranean, this worked well. But such a plow was inappropriate to the wet climate and often sticky soils of northern Europe. By the latter part of the 7th century after Christ, however, following obscure beginnings, certain northern peasants were using an entirely new kind of plow, equipped with a vertical knife to cut the line of the furrow, a horizontal share to slice under the sod, and a moldboard to turn it over. The friction of this plow with the soil was so great that it normally required not two but eight oxen. It attacked the land with such violence that cross-plowing was not needed, and fields tended to be shaped in long strips.

In the days of the scratch-plow, fields were distributed generally in units capable of supporting a single family. Subsistence farming was the presupposition. But no peasant owned eight oxen: to use the new and more efficient plow, peasants pooled their oxen to form large plow-teams, originally receiving (it would appear) plowed strips in proportion to their contribution. Thus, distribution of land was based no longer on the needs of a family but, rather, on the capacity of a power machine to till the earth. Man's relation to the soil was profoundly changed. Formerly man had been part of nature; now he was the exploiter of nature. Nowhere else in the world did farmers develop any analogous agricultural implement. Is it coincidence that modern technology, with its ruthlessness toward nature, has so largely been produced by descendants of these peasants of northern Europe?

This same exploitive attitude appears slightly before A.D. 830 in Western illustrated calendars. In older calendars the months were shown as passive personifications. The new Frankish calendars, which set the style for the Middle Ages, are very different: they show men coercing the world around them--plowing, harvesting, chopping trees, butchering pigs. Man and nature are two things, and man is master.

These novelties seem to be in harmony with larger intellectual patterns. What people do about their ecology depends on what they think about themselves in relation to things around them. Human ecology is deeply conditioned by beliefs about our nature and destiny--that is, by religion. To Western eyes this is very evident in, say, India or Ceylon. It is equally true of ourselves and of our medieval ancestors.

The victory of Christianity over paganism was the greatest psychic revolution in the history of our culture. It has become fashionable today to say that, for better or worse, we live in "the post-Christian age." Certainly the forms of our thinking and language have largely ceased to be Christian, but to my eye the substance often remains amazingly akin to that of the past. Our daily habits of action, for example, are dominated by an implicit faith in perpetual progress which was unknown either to Greco-Roman antiquity or to the Orient. It is rooted in, and is indefensible apart from, Judeo-Christian teleology. The fact that Communists share it merely helps to show what can be demonstrated on many other grounds: that Marxism, like Islam, is a Judeo-Christian heresy. We continue today to live, as we have lived for about 1700 years, very largely in a context of Christian axioms.

What did Christianity tell people about their relations with the environment?

While many of the world's mythologies provide stories of creation, Greco-Roman mythology was singularly incoherent in this respect. Like Aristotle, the intellectuals of the ancient West denied that the visible world had had a beginning. Indeed, the idea of a beginning was impossible in the framework of their cyclical notion of time. In sharp contrast, Christianity inherited from Judaism not only a concept of time as nonrepetitive and linear but also a striking story of creation. By gradual stages a loving and all-powerful God had created light and darkness, the heavenly bodies, the earth and all its plants, animals, birds, and fishes. Finally, God had created Adam and, as an afterthought, Eve to keep man from being lonely. Man named all the animals, thus establishing his dominance over them. God planned all of this explicitly for man's benefit and rule: no item in the physical creation had any purpose save to serve man's purposes. And, although man's body is made of clay, he is not simply part of nature: he is made in God's image.

Especially in its Western form, Christianity is the most anthropocentric religion the world has seen. As early as the 2nd century both Tertullian and Saint Irenaeus of Lyons were insisting that when God shaped Adam he was foreshadowing the image of the incarnate Christ, the Second Adam. Man shares, in great measure, God's transcendence of nature. Christianity, in absolute contrast to ancient paganism and Asia's religions (except, perhaps, Zoroastrianism), not only established a dualism of man and nature but also insisted that it is God's will that man exploit nature for his proper ends.

At the level of the common people this worked out in an interesting way. In

Antiquity every tree, every spring, every stream, every hill had its own genius loci, its guardian spirit. These spirits were accessible to men, but were very unlike men; centaurs, fauns, and mermaids show their ambivalence. Before one cut a tree, mined a mountain, or dammed a brook, it was important to placate the spirit in charge of that particular situation, and to keep it placated. By destroying pagan animism, Christianity made it possible to exploit nature in a mood of indifference to the feelings of natural objects.

It is often said that for animism the Church substituted the cult of saints. True; but the cult of saints is functionally quite different from animism. The saint is not in natural objects; he may have special shrines, but his citizenship is in heaven. Moreover, a saint is entirely a man; he can be approached in human terms. In addition to saints, Christianity of course also had angels and demons inherited from Judaism and perhaps, at one remove, from Zoroastrianism. But these were all as mobile as the saints themselves. The spirits in natural objects, which formerly had protected nature from man, evaporated. Man's effective monopoly on spirit in this world was confirmed, and the old inhibitions to the exploitation of nature crumbled.

When one speaks in such sweeping terms, a note of caution is in order. Christianity is a complex faith, and its consequences differ in differing contexts. What I have said may well apply to the medieval West, where in fact technology made spectacular advances. But the Greek East, a highly civilized realm of equal Christian devotion, seems to have produced no marked technological innovation after the late 7th century, when Greek fire was invented. The key to the contrast may perhaps be found in a difference in the tonality of piety and thought which students of comparative theology find between the Greek and the Latin Churches. The Greeks believed that sin was intellectual blindness, and that salvation was found in illumination, orthodoxy--that is, clear thinking. The Latins, on the other hand, felt that sin was moral evil, and that salvation was to be found in right conduct. Eastern theology has been intellectualist. Western theology has been voluntarist. The Greek saint contemplates; the Western saint acts. The implications of Christianity for the conquest of nature would emerge more easily in the Western atmosphere.

The Christian dogma of creation, which is found in the first clause of all the Creeds, has another meaning for our comprehension of today's ecologic crisis. By revelation, God had given man the Bible, the Book of Scripture. But since God has made nature, nature also must reveal the divine mentality. The religious study of nature for the better understanding of God was known as natural theology. In the early Church, and always in the Greek East, nature was conceived primarily as a symbolic system through which God speaks to men: the ant is a sermon to sluggards; rising flames are the symbol of the soul's aspiration. This view of nature was essentially artistic rather than scientific. While Byzantium preserved and copied great numbers of ancient Greek scientific texts, science as we conceive it could scarcely flourish in such an ambience.

However, in the Latin West by the early 13th century natural theology was following a very different bent. It was ceasing to be the decoding of the physical symbols of God's communication with man and was becoming the effort to understand God's mind by discovering how his creation operates. The rainbow was no longer simply a symbol of hope first sent to Noah after the Deluge: Robert Grosseteste, Friar Roger Bacon, and Theodoric of Freiberg produced startlingly sophisticated work on the optics of the rainbow, but they did it as a venture in religious understanding. From the 13th century onward, up to and including Leibnitz and Newton, every major scientist, in effect, explained his motivations in religious terms. Indeed, if Galileo had not been so expert an amateur theologian he would have got into far less trouble: the professionals resented his intrusion. And Newton seems to have regarded himself more as a theologian than as a scientist. It was not until the late 18th century that the hypothesis of God became unnecessary to many scientists.

It is often hard for the historian to judge, when men explain why they are doing what they want to do, whether they are offering real reasons or merely culturally acceptable reasons. The consistency with which scientists during the long formative centuries of Western science said that the task and the reward of the scientist was "to think God's thoughts after him" leads one to believe that this was their real motivation. If so, then modern Western science was cast in a matrix of Christian theology. The dynamism of religious devotion, shaped by the Judeo-Christian dogma of creation, gave it impetus.

AN ALTERNATIVE CHRISTIAN VIEW

We would seem to be headed toward conclusions unpalatable to many Christians. Since both <u>science</u> and <u>technology</u> are blessed words in our contemporary vocabulary, some may be happy at the notions, first, that, viewed historically, modern science is an extrapolation of natural theology and, second, that modern technology is at least partly to be explained as an Occidental, voluntarist realization of the Christian dogma of man's transcendence of, and rightful mastery over, nature. But, as we now recognize, somewhat over a century ago science and technology--hitherto quite separate activities--joined to give mankind powers which, to judge by many of the ecologic effects, are out of control. If so, Christianity bears a huge burden of guilt.

I personally doubt that disastrous ecologic backlash can be avoided simply by applying to our problems more science and more technology. Our science and technology have grown out of Christian attitudes toward man's relation to nature which are almost universally held not only by Christians and neo-Christians but also by those who fondly regard themselves as post-Christians. Despite Copernicus, all the cosmos rotates around our little globe. Despite Darwin, we are <u>not</u>, in our hearts, part of the natural process. We are superior to nature, contemptuous of it, willing to use it for our slightest whim. The Governor of California, like myself a churchman but less troubled than I, spoke for the Christian tradition when he said (as is alleged), "when you've seen one redwood tree, you've seen them all." To a Christian a tree can be no more than a physical fact. The whole concept of the sacred grove is alien to Christianity and to the ethos of the West. For nearly 2 millennia Christian missionaries have been chopping down sacred groves, which are idolatrous because they assume spirit in nature.

What we do about ecology depends on our ideas of the man-nature relationship. More science and more technology are not going to get us out of the present ecologic crisis until we find a new religion, or rethink our old one. The beatniks, who are the basic revolutionaries of our time, show a sound instinct in their affinity for Zen Buddhism, which conceives of the man-nature relationship as very nearly the mirror image of the Christian view. Zen, however, is as deeply conditioned by Asian history as Christianity is by the experience of the West, and I am dubious of its viability among us.

Possibly we should ponder the greatest radical in Christian history since Christ: Saint Francis of Assisi. The prime miracle of Saint Francis is the fact that he did not end at the stake, as many of his left-wing followers did. He was so clearly heretical that a General of the Franciscan Order, Saint Bonaventura, a great and perceptive Christian, tried to suppress the early accounts of Franciscanism. The key to an understanding of Francis is his belief in the virtue of humility--not merely for the individual but for man as a species. Francis tried to depose man from his monarchy over creation and set up a democracy of all God's creatures. With him the ant is no longer simply a homily for the lazy, flames a sign of the thrust of the soul toward union with God; now they are Brother Ant and Sister Fire, praising the Creator in their own ways as Brother Man does in his.

Later commentators have said that Francis preached to the birds as a rebuke to men who would not listen. The records do not read so: he urged the little birds to praise God, and in spiritual ecstasy they flapped their wings and chirped rejoicing. Legends of saints, especially the Irish saints, had long told of their dealings with animals but always, I believe, to show their human dominance over creatures. With Francis it is different. The land around Gubbio in the Apennines was being ravaged by a fierce wolf. Saint Francis, says the legend, talked to the wolf and persuaded him of the error of his ways. The wolf repented, died in the odor of sanctity, and was buried in consecrated ground.

What Sir Steven Ruciman calls "the Franciscan doctrine of the animal soul" was quickly stamped out. Quite possibly it was in part inspired, consciously or unconsciously, by the belief in reincarnation held by the Cathar heretics who at that time teemed in Italy and southern France, and who presumably had got it originally from India. It is significant that at just the same moment, about 1200, traces of metempsychosis are found also in western Judaism, in the Provencal Cabbala. But Francis held neither to transmigration of souls nor to pantheism. His view of nature and of man rested on a unique sort of pan-psychism of all things animate and inanimate, designed for the glorification of their transcendent Creator, who, in the ultimate gesture of cosmic humility, assumed flesh, lay helpless in a manger, and hung dying on a scaffold.

I am not suggesting that many contemporary Americans who are concerned about our ecologic crisis will be either able or willing to counsel with wolves or exhort birds. However, the present increasing disruption of the global environment is the product of a dynamic technology and science which were originating in the Western medieval world against which Saint Francis was rebelling in so original a way. Their growth cannot be understood historically apart from distinctive attitudes toward nature which are deeply grounded in Christian dogma. The fact that most people do not think of these attitudes as Christian is irrelevant. No new set of basic values has been accepted in our society to displace those of Christianity. Hence we shall continue to have a worsening ecologic crisis until we reject the Christian axiom that nature has no reason for existence save to serve man.

The greatest spiritual revolutionary in Western history, Saint Francis, proposed what he thought was an alternative Christian view of nature and man's relation to it: he tried to substitute the idea of the equality of all creatures, including man, for the idea of man's limitless rule of creation. He failed. Both our present science and our present technology are so tinctured with orthodox Christian arrogance toward nature that no solution for our ecologic crisis can be expected from them alone. Since the roots of our trouble are so largely religious, the remedy must also be essentially religious, whether we call it that or not. We must rethink and refeel our nature and destiny. The profoundly religious, but heretical, sense of the primitive Franciscans for the spiritual autonomy of all parts of nature may point a direction. I propose Francis as a patron saint for ecologists.

FUTURE IS FAMINE FOR TEN MILLION SOUTH OF SAHARA

      For all our scientific and technological prowess, we are still at the mercy of nature.

    The meteorologists' term is formal and dry: intertropical convergence zone. That is where the trade winds from the north and south meet in the vicinity of the equator, causing updrafts and clouds. Each year when the zone itself drifts northward over Africa, the rainy season begins along what is called the "Sahel," the southern border of the Sahara. From June to October the parched lands bloom and the nomads' cattle and goats grow fat.

    It may be that the air currents of the intertropical convergence zone lost some of their strength. Or perhaps the water temperature of the nearby Atlantic dropped slightly, lowering the moisture content of the low-level winds that feed into the intertropical convergence zone. Or perhaps there was a weakening of the much higher, faster easterly winds that blow over the continent, affecting Africa's rains and India's monsoons. In any event, some such comparatively small and maybe semi-permanent perturbation in the dynamics of the global climate has forced six African nations over the brink of catastrophe. It is by no means the first time within living memory, but for four years now--in Mali, Senegal, Mauritania, Niger, Upper Volta and Chad--the rains have failed.

    An embarrassment for the governments of these impoverished nations and a matter overlooked by a world with more noisome troubles, drought has silently crept into the lives of at least 30 million Africans. Nor are the troubles confined to Africa. Drought has almost circled the globe. Huge hungry mobs have rioted in

---

Copyright © 1973 by *Smithsonian* Magazine; reprinted by permission.

rainless areas of India; a combination of drought and floods has crippled food production in Bangladesh, Sri Lanka (formerly Ceylon) and many parts of Southeast Asia. The Philippines, this last year, produced 740,000 tons less rice than expected.

The miracle grains of the much vaunted Green Revolution--already under attack as oversold--cannot perform the ultimate miracle: They cannot grow without water. In the United States the labyrinths of diplomacy and economics have led to the exhaustion of our historic food surpluses. Just when most nations are living year by year, hand to mouth, drought has come to much of the Earth.

In West Africa the alarm was only recently sounded. Aid has begun to flow in along creaky and inadequate distribution channels and shaky statistics have begun to flow out: Of the nearly three million head of cattle in Upper Volta, only an estimated 500,000 remain; the human population of Mauritania's capital, Nouakchott, is now 120,000, swollen to three times its normal size as nomads drift in from the hinterlands to wait stoically for handouts; in parts of the region the desert itself is moving south at a rate of 30 miles a year.

But statistics, like meteorological terms, are dry and remote. They smack of reports and evoke a distant sympathy at best. Drought, however, is an intimate, personal disaster and this is what photographer Farrell Grehan documented along the northern banks of the Senegal River, normally Mauritania's greenest area.

As he set out from Nouakchott to the small river town of Rosso late last spring, Grehan looked for dead cattle along the road. "We did not wait long. Flies covered the carcasses of cows, donkeys, goats and even camels. Excited and appalled, I kept taking the same kind of picture. Getting the best angle placed me to leeward as well as to windward of the carcasses. I had to gulp for air, hold my breath, focus on the hollow eyes with eyelashes of blue flies and then do it all over again from a different angle."

In recent years, immunization campaigns and newly dug water holes permitted the herds to increase. The resulting overgrazing was then compounded by the failure of the rains. Forage began to disappear. Water holes began to dry up. Livestock began dying and the nomads moved their dwindling herds southward. More than 35 percent of Mauritania's cattle have died; some 40 percent have been moved to Mali and Senegal and farther south, straining the resources of these already straitened countries.

"I approached a herd of a dozen cattle resting motionless on the ground," wrote Grehan. "Finally they started to rise, moving one limb at a time agonizingly. It was as if a dignitary paid an unexpected visit to an old people's home; everyone tried to get up but no one could summon the strength. Later, in Boghe on the banks of the Senegal River, I saw two cows chewing with intensity and concentration. One was attempting to eat a corrugated cardboard box, the other a burlap bag, all in dreadful slow motion. Only once did I see an animal being fed. It was a scrawny horse that must have been famished but it too munched mechanically like a children's toy winding down."

Unlike a toy that winds down, the conditions of drought tend to be self-perpetuating. The soil dries up, the surface grows hotter and the relative humidity therefore is lower, impeding cloud formation. Furthermore, for whatever moisture there is in the atmosphere to condense into clouds, "condensation nuclei" are needed --small particles given up to the atmosphere by the land's surface. In drought-ridden West Africa the land offers only dust, such tiny particles of dust that even when water does condense on them they do not grow heavy enough to fall from the sky.

Grehan was overtaken by a dust storm on the road from Rosso to Boghe. "In many places," he wrote, "windblown dust covered everything. It stung the eyes, filtered through clothes, covered the body's pores and daubed every black, brown or white man the same dun-colored yellow. Here were grasslands without a blade of grass:

once fertile soil flung into the air, extinguishing the African sun."

And so the drought continues. Throughout the desert, thousands of herdsmen have tried to save their livestock by feeding them the foliage from the desert's trees. With trees cut down and shrubs devoured, there is little vegetation to keep the soil in place. Thus the desert grows. Creating its own weather, pushed by overgrazing and impelled by the action of desperate men, the Sahara continues south as it has for centuries--but at an accelerated pace. It has leapfrogged the Senegal River, once a natural barrier.

In yet another way the drought feeds upon itself. In many areas, the drought prevented farmers from planting this year's crops; in other areas, harvest fell short. Having little or no food reserves, the farmers have been driven to eat their crop seed, thus eliminating the possibility of planting next year's crops.

"While livestock deaths are appalling," reported Jerry E. Rosenthal of the U.S. Agency for International Development, "crop losses throughout the Sahelian Zone are no less devastating. Grain production ... has dropped 50 percent in many areas of the region. And in Mederdra, Mauritania, the fields which can be cultivated grow smaller each year as desert sand takes over. A forlorn village chief says: 'I am witnessing the burial of my village.'"

The desert grows, the crops fail and villages are abandoned--how many no one knows. Grehan spent time in one such deserted place: "Only the bare skeleton of the village was standing, a few wooden bowls and a ladle or two left behind, giving rise to the question of whether the people departed in great haste or had hopes of returning one day. Where could they go? To Mali? Senegal? Presumably those ravaged lands cannot be too happy about adding outsiders' problems to their own. Can these tribesmen find any land in all of West Africa which could possibly support them?"

The answer seems to be no. No place could support all of them. They flock to the towns and cities and to the rivers, seeking water, food and medical help. Jerry Rosenthal reported that Timbuktu's harbor on the Niger River has dried up to the point that barges can no longer reach the city. But there is water in the river and the nomads, whose families have always eaten meat and milk, wait hours in the 120-degree heat for a handful of grain.

Like the Niger, the Senegal River, one of Africa's largest, has dwindled to a trickle. Grehan noted that "one could walk across the river and in most places the water would barely reach above one's knees." The diminished rivers teem with people, carrying on much as they would in normal times, hoping--cheerfully assuming--it would rain soon.

That the rivers were drying up, that the drought would be serious dawned on officials well over a year after it began. (Of the six Sahelian countries, four are listed euphemistically by the United Nations as "the least developed of the developing nations." In spite of high mortality, their populations are growing. Literacy averages about ten percent and communications, particularly in the desert areas, are limited to the point of virtual nonexistence.) Late in 1971, the Food and Agriculture Organization's early-warning system began issuing ominous reports about local rain and crop failures, and at about the same time AID personnel on the scene had begun to sound the alarm. In 1972 food shipments--for both livestock and humans-- started to trickle in and several Sahelian countries inaugurated rural development programs and well-digging efforts. Not until the end of March of this year did the six countries declare their region a disaster area and make an all-out appeal for help.

By early June animal feed and seeds were being airlifted into the region under FAO auspices. (Not just any seeds will do. They must be adapted to the climate and other conditions of the area; optimally the seeds should come from the Sudan and other nearby countries but Sudanese surpluses were insufficient, so much of next year's crop may fail even if it does rain.) In the United States various private organizations, including Operation Push headed by Jesse Jackson, began to make appeals for food, and AID, though strapped by national economy strictures, nonetheless committed $24 million for food grain (156,000 tons) and its transportation-- theoretically enough to feed 11 million people a pound of food a day for one month. By mid-June American air transports were aiding in relief operations, along with planes from other nations, but late that month a grateful Andre Coulbary, Senegal's ambassador to the United States, publicly worried that the American grain, plus some 250,000 tons from other nations, would be too little and would arrive too late. Ironically, if the prayed-for rains do come, roads would be rendered impassable and the food could not be distributed. Even without rains large sections of the region, particularly in Upper Volta, are beyond the reach of trucks and planes.

The nightmare that haunts West Africa now is, of course, famine. Most of the frantic relief efforts that gained momentum over the summer were mounted simply to avoid the starvation of some six million people. And by mid-July, experts expected that by October ten million people would be threatened with starvation. By July there had been no deaths reported with starvation as the cause. But thousands had died--no one knew how many. Was it that a famine had begun and no one would admit it?

Famine is so horrendous a part of human experience that no one wants to think about it. Officials and experts are leery of using the word lest they fall into the hazard of crying wolf. Furthermore, health officials generally list deaths by their final, specific cause rather than record a generalized condition, such as malnutrition, that may lead to a specific disease. But if widespread malnutrition resulting in death is famine, then there was famine in West Africa as early as last spring.

AID's Rosenthal reports that unknown numbers of children and old people are dying of measles and other diseases to which their underfed systems have little or no resistance. And of course what would be a minor threat in a highly developed country can be a disaster elsewhere: Medical facilities in the impoverished Sahelian Zone are minimal.

Farrell Grehan watched people lining up outside the offices of Boghe's only clinic. Operated by Algerians, it was indeed the only clinic for hundreds of miles around. "Two camels loped into the courtyard's center and were made to settle down on the ground. I watched the elegance of the animals and the delicate movements of the nomads as they loosened a small bundle from the first camel's back and carefully lowered it to the ground.

"It turned out to contain a very tiny, very lovely young woman, heavy silver jewelry making her frail arms seem even frailer. She had a baby with her. The nomads carried her in to the doctor's office where he applied a stethoscope and made his diagnosis. His assistant gave her shots and, since there were no hospital facilities, she was put on the bare floor in the end of the clinic where two young men and an old woman already lay with the look of death in their eyes.

"During the second night, the tiny woman died. The next morning I saw her child sitting alone in the clinic so covered with flies around his mouth and chin that he appeared to have a beard."

PROPHECIES AND HARD QUESTIONS

As the summer drew to an end without rain, the immediate question was: How many more people will die, will there be a full-scale famine despite the best efforts of the world to head it off? Beyond that--once the drought ended--lay some further hard questions and some awful knowledge. It was known, for example, that thousands, probably millions, of young Africans would grow up blighted in their infancy by malnutrition, retarded in mind and body. It was known also that it would take at least three to four years for a tribe that had lost its livestock to build up a herd and regain self-sufficiency.

Would the tribesmen settle for smaller herds made up of better cattle (if indeed that could be arranged), thus perhaps avoiding the overgrazing that, even without a drought, had been destroying the land? Where would the money, the energy and expertise come from to make better use of the region's rivers for irrigation and to find dependable sources of underground water? How would these destitute countries raise taxes, increase literacy rates, train and deploy medical personnel, control their population growth?

And amid all these questions one other seemed worth pondering. About ten years ago a number of scholars predicted darkly that given the size and present rate of growth of the population, the human race would inevitably experience a series of severe famines and social and political disruptions. Some ventured to say that the troubles would begin in earnest as early as 1975. So as the summer of 1973 ended and thousands of hungry Africans meditated on the outskirts of dusty villages and what seemed like a number of unrelated crop failures, dry spells and floods plagued the midriff of the planet, one wondered if the prophecies of those quickly dismissed as Neo-Malthusian alarmists should not now be reconsidered.

*

A-BOMBS, BUGBOMBS, AND US
G.M. Woodwell, W.M. Malcolm, R.H. Whittaker

>   More insideous than the problems caused by overpopulation
> are the results of environmental pollution. As the following
> series of articles indicates, pollution need not be obvious to
> be threatening.

   THE WORLD IS SMALL. This is the lesson of the Bomb and the rockets, of hunger and of exploding populations. And it is the lesson that the problems of POLLUTION also teach. We once thought that dilution of man's wastes into the earth's vast currents of air and water was the simple answer to all problems of waste disposal. We know now that these currents are not vast enough to handle safely all the wastes and poisons man is releasing into them.
   Two aspects of environmental pollution--radioactivity and pesticides--illustrate the problem most effectively.
   The lessons from these call for restraint and a gradual revolution in the use of environment and in pest control. It is these twinned problems that are our subject.
   Two ecological ideas are at the heart of these pollution problems: First is the principle that substances released into environment move in pathways loosely described as "cycles" and often return, concentrated, to threaten man himself. Second, the poisons used to control pests have effects on many populations, not merely the pests; effects of these poisons include: (a) killing of some wild animal populations, especially those of predatory animals which regulate populations of other animals; (b) causing population erruptions of other species, which may become new pests, while (c) the old pests remain and evolve new ability to survive the poisons.

Brookhaven National Laboratory Publication No. 9842, 1966. Reprinted with permission.

Much of what we know about biological, geological and chemical cycles of the earth has come from studies of radioactive tracers, and especially fallout from bombs. What these studies have taught us is closely related to what we have been hearing more recently about movement of pesticides. For this reason a brief review of the lessons of radioactivity is appropriate.

Biologists have long known that substances are carried from place to place in environment by currents of moving water, air and by moving organisms, and that substances also are transferred from one kind of organism to others. It took an unfortunate series of mishaps in the Pacific in the mid-1950's to demonstrate the meaning of such cycling for man. These events brought rapid quickening of interest in environment and even forced governments to examine these questions.

The bomb test at Bikini in 1954, known as BRAVO, due to unexpected winds at upper levels dropped radioactive fallout on Rongelap Atoll and exposed the inhabitants including several U.S. servicemen to significant radiation (8, 10). A Japanese fishing vessel was also in the fallout field and its crew was exposed as well. At the time that these events were being reported in the world press, fish in Japanese markets were discovered to contain fallout radioactivity in easily measurable, if not genuinely hazardous, amounts. Public reaction was rapid and intense, especially in Japan. A Japanese oceanographic vessel and later an American one, were sent on large sweeps through the Pacific and found that the contamination was considerable and was widespread (10). Wind and water had spread the radioactive material from this and earlier tests across wide areas of the Pacific. In each area that radioactive dust fell into the ocean, the material was taken up by small plants, the plants were eaten by small animals, the animals were eaten by other animals, and these in turn were eaten by such large animals as tuna fish, which accumulated the radioactive material passed up to them along the food chain. Migrating tuna fish could carry their radioactivity considerable distances, even to the last link in the chain, man.

Although the real hazard to human health was small, the Japanese, who depend heavily on food from the sea, were understandably alarmed. Most people had been acutely aware for years of the problem of direct radioactive fallout. The hazard to man from cycling in currents of wind and water and in organisms was a less obvious, disturbing discovery. The incident gave dramatic proof that there are limits to the capacity of wind and water to dilute pollutants potentially harmful to man.

Since then there have been many efforts to clarify the biochemical cycles of the earth using isotopes from fallout (2, 6, 12, 14). Most of the knowledge of these cycles, fragmentary though it is, has been gleaned from researches started since 1954 by a very large program of environmental research within the AEC costing many millions of dollars.

There is no simplified general rule describing the movement of radioactive materials in the environment. Each chemical element and each substance travels its own peculiar path, largely independent of pathways of other elements. Time and place of release, size of particles, chemical composition, environmental conditions, and kinds of organisms present are all important in determining patterns of movement. While the substances released tend first to be diluted in the water, air, or soil, they are frequently concentrated again later by organisms, often in unexpected places. It is often the case that a substance which is very dilute in the water is picked up by algae and transferred along food chains at concentrations many thousands of times as high as those in the water (3, 5, 13, 18). The tendency to move (even substances normally considered "insoluble" move) and to be concentrated presents the dual hazard of wide dispersal and potentially hazardous local concentrations of pollutants. This means that there is continuing and expanding need for rigorous checking of environmental concentrations, and especially of man's own food chains for a wide variety of pollutants. From the studies of radioactivity we know that the "cycles" of the biosphere in which pollutants travel are too complex for easy or safe prediction, and that we cannot simply trust dilution to protect man against the persistent poisons he releases into environment.

With radioactivity, pesticides, and other poisons, the principal direct hazard to man arises from contamination of food chains that man taps. A few of these have been studied in detail. Radioactive iodine (iodine ^{131}) for instance is a particularly serious hazard because it emits gamma rays and is strongly concentrated in the thyroid gland of man and other mammals. Iodine ^{131} has an 8-day half-life and is a hazard only during the first weeks following its release, but in those weeks it may reach high levels in man even though the contamination in environment was low. It enters several of man's food chains but the most important route is through cow's milk. The food chain in this instance is short and simple--involving only grass, cattle, milk and man--but it is difficult to predict the extent of iodine ^{131} concentration into man's thyroid gland. The rate of movement to man is affected by (a) the density of the grass the cow eats (if the forage grows densely and the cow covers little ground, she picks up less iodine), (b) rain, which may wash some of the radioactive particles off the forage into the soil, (c) the diet of the cow (she may be eating heavy supplements of grain or stored hay), (d) the efficiency with which iodine is secreted in the milk, and (e) the amount of milk consumed by individuals who may differ greatly in their consumption, and other factors. It is not easy, first, to predict the concentration into man or, second, to predict the possible long-range meaning to the human body of a given concentration. There is another lesson from these studies--they are full of surprises, sometimes tragic ones. One such surprise of recent years was discovery of tumors of the thyroid gland in children caused by radiation exposures at levels that had been assumed to be safe. The effect was discovered in studies of children whose thyroids were irradiated when they were treated with radiation years ago for enlarged thymus glands. Recognition that the thyroid may be damaged after relatively low exposure to radiation (either by x rays or from the emissions of iodine ^{131}) led to issuance of the Radiation Protection Guide by the Federal Radiation Council in 1961, setting limits at less than 1/10 previously acceptable levels (7). The wisdom of this step was demonstrated when studies of the children exposed at Rongelap Atoll in 1954 showed occurrence in them of thyroid tumors--tumors that did not appear until about 10 years after irradiation ...

These experiences involved an isotope whose environmental pathway to man is relatively simple and whose clinical effects seem comparatively clear cut. Yet even here, where biologists had reasonable assurance that they knew the mechanisms, we have been surprised. (It may well be that only <u>because</u> we know the mechanisms well were we able to recognize the problem.) The lesson is an important one: apparently even when we think we know the pathways and the concentrations that are safe, there is reason for conservatism in estimating hazards to man. When hazards are less clearly known, as is the case with pesticides, there is greater cause for caution. It is unusual to be able to link a pollution cause to a health effect ten years later. Such effects could be established for radioactive materials only because of a very large body of background data on radiation effects and because of an unusual and unequivocal, if unfortunate, series of radiation exposures. It may be many years before equivalent knowledge of pesticides in relation to environment and man's health is available. Pesticides are intensely poisonous, and must be poisonous to control pests effectively. There is little evidence now that they produce cancer or are (except for special cases of extreme exposure) damaging to health. But it is certainly not true that hazard to man is unlikely. It is more accurate to say, first, that we do not know now the long-range hazard of pesticides to man and, second, that analogies with radioisotopes and cancer-producing substances in cigarette smoke and atmospheric pollution suggest we are likely to encounter unpleasant surprises.

There is one more important point about food chains as they affect both radioactive materials and pesticides. This is the fact that there is often a step-by-step increase in concentration along a food chain. In the step from pasture to milk in our example of I^{131}, the cow gathers fallout from many square yards of grass and deposits it in 10-20 quarts of milk daily. In the next step iodine from a quart of milk daily is concentrated in the small volume of a child's thyroid gland. With other isotopes and in other food chains parallel patterns of increasing concentration are common. Greatest concentrations often occur in the last links of the chain, which may be man and the animals he eats. Generalizations, however, are difficult because the behavior of each element must be considered individually before its hazard to man can be appraised. The best survey of this work with radionuclides is the 1962 UNSCEAR Report and its 1964 supplement (15, 16).

No such authoritative document outlining hazards to man is available for pesticides, although the problem is an even more difficult one. Pesticides, which emit no easily measured signal similar to a radioactivity, are very difficult to detect. Further, the significance of pesticides as contaminants of the biosphere has been recognized only recently, and broad-scale research on the cycling of pesticides has really only barely started. Indeed, the greatest steps have been made in this work since 1960 when a very sensitive analysis technique for pesticides (gas chromatography) first became widely available for this application. And even now our knowledge is largely restricted to the persistent chlorinated hydrocarbons, especially DDT.

There are a few instances where studies of food chains have given indications of step-by-step increase in pesticide concentrations along food chains. For instance, in the food web in which the herring gull is a scavenger in Lake Michigan, DDT (DDE and DDD) concentrations in the bottom muds at 33-96 feet averaged 0.014 parts per million. In a shrimp (Pontoporeia affinis) they were 0.44 ppm, more than ten times higher. Levels increased in fish to the range of a few ppm (3.3 alewife; 4.5 chub; 5.6 whitefish) another tenfold increase, and jumped in the scavenging, omnivorous herring gull to 98.8 ppm, twenty times higher still, and 7000 times as high as in the mud (9). There are numerous less complete examples of this type of concentration of pesticides, especially the chlorinated hydrocarbons in birds (17, 20, 21).

This food chain effect is especially serious among birds, and predatory birds (hawks, owls, ospreys, and the eagle, our national emblem) are in real danger of extinction in the United States. Accumulation of pesticides can kill adult birds directly, as has repeatedly been observed in robins and other species in areas sprayed with DDT. The more serious effect, however, is a less obvious one--lower levels of pesticide accumulation appear to cause the birds to fail to reproduce successfully. The peregrine falcon, for instance, no longer breeds in the Northeastern United States, although there were numerous breeding sites there a decade ago (17). The Northeastern strains appear to have been wiped out, probably an example of pesticide extinction of a wild species over a wide area. Populations of ospreys or fish hawks and eagles appear also to be declining seriously.

It is a widely accepted ecological principle that the stability of populations is related in complex ways to the number of different kinds of organisms present. The greater the number of kinds of organisms (diversity), the greater the stability. In somewhat oversimplified terms diversity is thought to introduce stability by including predators and competitors of each species, thereby guaranteeing natural controls of population size. Reducing the number of species may remove predators or competitors, and allow spectacular increases in the populations that survive. The most abrupt changes occur in species that have short life cycles. Insects are among the most conspicuous of these rapidly reproducing populations. Populations of mites often erupt in farm fields after spraying with DDT, and there are numerous other examples (4). The principal point about pesticides is that control of pests with broad-spectrum poisons not only reduces the pest population but also reduces its competitors and predators. This decreases the stability of the populations, increasing the probability that populations of other resistant species may themselves reach "pest" proportions. Furthermore, this removal of predators and competitors implies that the species against which the pesticide was used is even more likely to erupt out of control after the spraying stops, than it was before. Pesticide spraying thus tends to intensify the pest problems the spraying was intended to control. Once pesticide use has begun, continued or increased use of pesticides is likely to be needed to control population instability which was in part produced by the pesticides.

But short life cycles mean something else--capacity for rapid evolutionary change. When fields are sprayed with pesticides, the individuals of the pest species that are most resistant are the ones that survive. They are also the ones that reproduce, and the population of the pest in that field the next year consists of their descendents which are, on the average, more resistant to the pesticide. Year by year the pest population becomes more resistant to the pesticide. There is consequently need either for heavier use of the pesticide, or for different kinds of pesticides to control its population. Meanwhile, new pest species may be appearing. Use of pesticides is thus a double-edged sword. On one side it has great short-range advantages for simple, effective control of farm pests. On the other, long-range side, it is a kind of treadmill, or a constantly escalating chemical warfare. Because it creates new pest problems and pest resistance, there is need for constantly increasing amounts and variety of pesticides to control pests.

If it were not for the cycling effects, this warfare might be accepted without concern. The observations on birds suggest, however, that steadily increasing use of pesticides will cause extinction of increasing numbers of more vulnerable wild species. Furthermore, the cycling effects imply increasing consumption of pesticides in food by man.

Two ominous facts should be placed side by side. First, is the rapid increase in pesticide use. Money spent for pesticides in the U.S. increased between 1962 and 1964 from about $734 million to $944 million (it is encouraging that DDT use in this period declined; most of the increase was in organic phosphate insecticides) (19). Second is the fact that certain pesticides such as DDT can now be found in tissues of animals from the sea. The problem works this way:
1) DDT and some other pesticides are stable chemicals. They do not simply break down into harmless forms in the soil or water, but tend to accumulate there in increasing amounts.
2) Rapidly increasing pesticide use implies increasing amounts of pesticides accumulated in such reservoirs.
3) From farm soils they may be carried as dust (1) and by moving water into streams, where they kill some fresh water life. (Devastating kills of millions of fish in the Mississippi appear to have resulted from careless contamination of that river by endrin, one of the pesticides. The deleterious effects of DDT spraying for the spruce budworm on salmon runs in the Miramichi River in New Brunswick, Canada, have been very carefully documented) (11).
4) Rivers and winds carry the pesticides into the ocean. There is a steady move-

ment of pesticides from farm fields through streams into the ocean, and consequent increasing accumulation of pesticides there.

5) In the ocean pesticides are concentrated by food chains into animals. Man eats some of the animals which are last links of food chains.

6) Gradually increasing accumulation of pesticides from food into human tissue is implied. Man is also receiving some pesticides in food he eats that was grown on land.

7) So far as the ocean is concerned, some nations are now dependent on it for an important part of their food; and as man's population grows, this dependence will increase.

Even the oceans are not so large that man can assume he will not, because of dilution effects, be affected by pesticide cycling. As we have indicated, the long-range meaning to the human body of a variety of different pesticides accumulated in its tissues is not known. Use of pesticides has great and compelling short-range, practical advantage. It is unfortunately also true that pesticide use has marked long-range disadvantage in its effects on natural population and pest evolution and (probably, with due allowance for what we cannot now know) on human health.

Steadily increasing contamination of environment with radioactive materials from bomb fallout would be, in long-range terms, foolish and reckless. This is one reason for the treaty banning tests of atomic bombs in the atmosphere. Uncontrolled, constantly increasing contamination of environment by persistent pesticides would be, in long-range terms, equally foolish. Yet use of some pesticides is almost essential as a part of the means of controlling pests. What can we do? These steps seem necessary:

1) Restriction of the persistent, broad-spectrum poisons such as the chlorinated hydrocarbon insecticides to uses that cannot contaminate the biological, geological and chemical cycles of the earth.

2) Replacement of these pesticides by others that are not persistent (and whose breakdown products are not toxic) and are specific (kill only the pest they are aimed at). There is need for extensive research in developing such pesticides.

3) Above all, learn to use pesticides with restraint as only one tool in a wide range of possible control techniques that include use of a diversity of resistant crop strains, use of natural enemies to control pests, and use of a greater diversity of crops in any area. Maintenance of the full range of species in the natural communities of an area is probably one important step toward this shift in pest control practices. Research on the implications of biological diversity (number of species per unit area) for controlling pests is being pressed at present and will doubtless contribute new techniques enabling the transition.

More broadly, we must restrain the use of pesticides for the sake of our own future. The world is now too small for rapidly increasing human populations to release rapidly increasing amounts of persistent poisons into environment. Man needs to begin a quiet revolution in his thinking, which recognizes the crowding of our world and seeks to govern man's own effects on environment with foresight, long-range wisdom, and restraint. In an age when we can reach for the moon, such a revolution seems possible. There is no doubt that it is necessary.

GENERAL BIBLIOGRAPHY

For detailed and authoritative discussions of the "Pesticides Problem" see:

Carson, R., <u>Silent Spring</u>. Boston: Houghton Mifflin, 1962.

Carson, R., Statement before the Subcommittee on Reorganization and International Organizations of the Committee on Government Operations, U.S. Senate, Washington, D.C., June 4, 1963. Part I, pp. 207-12, 1963.

DeBach, P., *Biological Control of Insect Pests and Weeds*. New York: Reinhold, 1964.

Rudd, R. L., *Pesticides and the Living Landscape*, Madison: University of Wisconsin Press, 1964.

*

SHIPPINGPORT — THE KILLER REACTOR
Richard Lewis

Doubts about the safety of nuclear power in the United States have been intensified by the claims made by Ernest J. Sternglass, long-time antagonist in the reactor debate, that the Shippingport reactor is releasing radioactive material far in excess of legal limits.

A fact-finding committee of doctors and scientists appointed by Gov. Milton Shapp of Pennsylvania has opened an inquiry into charges that radioactive emissions from the 90 megawatt nuclear power station of the Duquesne Light Co. at Shippingport, Pa. have resulted in a sharp rise in infant mortality and cancer deaths in adjacent areas of western Pennsylvania and eastern Ohio.

The Shippingport station is on the Ohio River 35 miles west of Pittsburgh. It is a pressurised water nuclear reactor of the type developed for nuclear submarines. Constructed jointly by the US Atomic Energy Commission and the utility as a demonstration plant in 1957, Shippingport has the historic distinction of being the first commercial nuclear power plant in the United States. After 16 years, it has become the first to be investigated by a state commission as a threat to public health.

The charges that it is indeed a threat were brought by Ernest J. Sternglass, a radiological physicist at the University of Pittsburgh, and a longtime critic of the AEC's radiation policies. For more than 10 years, Sternglass has tried to prove a correlation between low level radiation—from reactors, bomb fallout or any other source—and the incidence of fetal and infant mortality, childhood leukemia and

From *New Scientist*, September 6, 1973; used with permission.

cancers in all age groups. His claims that leaking reactors have increased infant deaths in the vicinities of power plants in Northern Illinois, Northern Michigan, and at three sites in New York State have been brushed off consistently by local, state and federal health officials and by some, but not all, of his peers in the scientific community.

Late last year, the indefatigable Sternglass seemingly hit the jackpot. He dug up a radiological monitoring survey of Beaver County where the Shippingport plant is located showing anomalously high concentrations in 1971-72 of iodine 131 and strontium 90 in the soil and in milk produced at dairies in the area. Curiously enough, reports of radioactive emissions from the plant and by the utility by the AEC during this period did not reveal any unusually high discharges.

But the anomalies in the surrounding soil and nearby milkshed could only have originated at Shippingport, Sternglass testified before a public hearing of the Governor's fact-finding committee on 31 July. Early in 1972, he said, there was an upsurge of radio-iodine. Levels in milk produced by dairies within 10 miles of Shippingport reached concentrations of 121 picocuries per litre--21 per cent above the federal maximum permissible level of 100. Between 19 March and 1 April, 1971 a dose rate of 410 millirems a year was measured on the plant site, approaching the AEC's legal maximum of 500 mr per year at the plant fence. Outside the plant boundary, dose rates of 371 mr per year were recorded in March 1971 for the community of Shippingport, followed by peaks of 352 mr in May, 292 mr in July, and 306 in November 1971. The annual average dose for Shippingport that year was 180 mr, or 96 above the natural background reading of 86 mr. The additional dose of 96 mr to the population of the community was 56 per cent of the maximum permissible annual dose of 170 mr to an individual in a population--a standard now widely regarded as unsafe in the United States.

The radionuclide concentrations were reported in a radiological survey by Nuclear Utilities Services, a private monitoring firm in Maryland, for Duquesne Light Co. The report was part of an environmental impact study the utility was required to submit to the AEC with an application to build two more nuclear power generating units at Shippingport.

The high levels revealed in the survey should have rung alarms throughout the entire public health establishment and particularly in the AEC and the Environmental Protection Agency. But as later testimony unfolded during the two-day public hearing of the gubernatorial committee, Sternglass was the only one who became concerned.

During the spring of 1971, levels of radioactivity in Ohio River bottom sediment rose in parallel to the upsurge of strontium 90 in the soil and milk of the area. The peak in the river sediments was 10 times the 1959 level.

VITAL STATISTICS

Looking at the offical vital statistics for the region, Sternglass reported that increases in infant and in cancer mortality seemed to respond to the increases in radionuclides in 1971-72. This pattern, he said, was historic. It could be traced back to 1958, the year after the Shippingport plant went into operation, and was particularly noticeable in other years when upsurges in radioactive emissions were reported.

He testified that radioactive waste releases from Shippingport had peaked in the period 1960-64. Four to six years later, mortality rates for all cancers in predominantly rural Beaver County (population 200,000) rose 39 per cent--from an annual rate of 148 per 100,000 in 1958 to 205 in 1968. In the same period, the vital statistics showed an increase in cancer of only 9 per cent.

Further, the cancer rates of other communities rose inversely with distance from Shippingport. At the town of Midland, 1-3 miles downstream on the Ohio from Shippingport, cancer mortality rates rose 184 per cent from the period of 1959-61 to 1970. In 1964 and 1970, the gross beta radioactivity in the river at Midland exceeded permissible discharge limits by 130 to 300 times, Sternglass said, citing official state reports.

During two 12-month periods, 1964-65 and 1970, the reactor dumped 78 and 182 curies of radioactivity respectively into the river, Sternglass said. Under terms of a state industrial waste permit issued in 1957, the reactor was allowed to discharge only 0.58 curies a year into the river. Later testimony at the hearing disclosed that in 1964 the Duquesne Light Co. asked the state to permit a tenfold increase in the discharge limit and the request was granted.

West of Shippingport and Midland, the river flows into the State of Ohio and it carried its radioactive burden there. Between 1958 and 1971, the cancer death rate rose 67 per cent in East Liverpool, Ohio, while the rate for the rest of the state increased only 4 per cent. Sternglass pointed out that residents of East Liverpool obtain their domestic water supplies from the river. Outside the city, however, residents get their water from wells and other sources. As a result, he said, the cancer death rate in adjacent areas outside the city is slightly lower than the state average.

Downstream from East Liverpool, the city of Steuvenville, some 30 miles west of Shippingport, showed an increase in cancer deaths of 25 per cent in the period of 1958-1971.

Nine miles east of Shippingport, the Borough of Aliquippa (population 23,000) exhibited an infant mortality rate of 16 per 1000 live births between 1949 and 1952. This was a period of extreme air pollution from the steel mills there. But in 1970, even after air pollution had been greatly reduced, the borough's infant mortality rate jumped to 43.9 per 1000 live births--compared with 20.2 for the rest of the state. In 1971, the rate dropped to 39.7 in the borough and to 18.3 in the state, but the fetal mortality rate was 55.7 per 1000 births, compared to 22 for the state.

THE AEC'S WRONG LIMITS

A pattern of similar rises in childhood leukemia, adult lung cancer, and even heart disease could be tied to increases in radioactive emissions from the pioneer Shippingport reactor, Sternglass asserted. He pleaded for a full, official investigation of health effects of the Shippingport and other reactors. It would require an intensive epidemiological study of nuclear power plants in Pennsylvania. For the first time, he won support in this plea. Another witness, Irwin D. J. Bross, director of biostatistics at the prestigious Roswell Park Memorial Institute, Buffalo, NY, also called for health surveillance systems in the vicinity of nuclear power plants.

Bross testified that the AEC's method of setting gross radiation exposure limits for a population was wrong because it ignored the demonstrated existence of a group of persons who are highly susceptible to radiation effects--such as leukemia and other cancers. The Institute's widely studied "Tri-State Survey" to determine background causes of leukemia had shown an astonishing difference between susceptible and non-susceptible groups, Bross said. Diagnostic X-rays which produce no discernible effect in non-susceptibles can result in an increased cancer risk of 1000 per cent in susceptible persons, the study revealed. This means that susceptible persons may be vulnerable to dosages an order of magnitude or two lower than the "safe" dose for non-susceptible persons, he said.

ERRONEOUS REPORT

Rebuttal to Sternglass's charges from the nuclear industry and from official agencies took a surprising turn. The Nuclear Utility Services report of high radionuclide levels in soil and milk at Shippingport was repudiated by the company's management. Joseph DiNunno, a former member of the AEC's regulatory staff and vice-president of NUS said the data had been re-checked at the request of the client, Duquesne Light Co., and appeared to be erroneous. NUS had asked four independent laboratories to re-survey the area, he said.

The Environmental Protection Agency, which has assumed radiological monitoring duties from the AEC, reviewed the NUS data and concluded they must be erroneous, deputy EPA administrator, William D. Rowe, told the committee. Shippingport is a very clean plant," said Rowe. He rejected the idea that any epidemiological study was necessary to determine health effects.

The EPA dictum, however, was contradicted by Irving Michelson, who appeared for Consumers Union. Analysis of US Public Health Service reports of radionuclide contamination of milk and other foodstuffs produced and sold in the Shippingport region "clearly points to excessive releases from Shippingport during the years 1970-71," he said, "and unless there is clear evidence to the contrary there is cause for real concern."

The Pennsylvania Department of Health, however, did not share that view. Dr. George K. Tokuhata, director of its Bureau of Program Evaluation, reported that although the infant death rate in Aliquippa in 1971 was 2-1 times that of the state, "we conclude that the infant mortality rate in Aliquippa is no higher than expected." He said a large proportion of nonwhite and illegitimate births in the steel making community "can readily explain the observed relatively high levels of fetal, neonatal and infant mortalities."

Dr. Gerald A. Drake, a physician specialising in internal medicine at Petoskey, Michigan, told the committee about a study he made of the health effects of a 75 megawatt boiling water reactor power plant at Big Rock Point in Charlevoix County, a rural resort area on Lake Michigan. Biota sampled near the reactor were 2 to 5.1 times more radioactive than those sampled five miles north or south. In 1971, infant mortality in the county was 49 per cent higher than the state average; leukemia deaths were 400 per cent above the state average and congenital defects 230 per cent above, he said.

The fact-finding committee is to issue its report to Governor Shapp this fall.

RICHARD LEWIS is editor of Science and Public Affairs--Bulletin of the Atomic Scientists

SUNLIGHT AND THE SST

Despite the pleas of the musical <u>Hair</u> to "let the sunshine in," any beachgoer knows empirically that too much sunlight hurts. During the dispute over the supersonic transport (SST) in 1971, it was suggested that a fleet of SSTs might reduce the amount of ozone in the stratosphere, letting too much sunshine in. Amid Washington rumors of an SST resurrection, an ad hoc panel of the National Academy of Sciences has now completed a report voicing the "utmost concern over the possible detrimental effects on our environment by the operation of large numbers of supersonic aircraft."

The cause of this anxiety is ultraviolet radiation (UV), that portion of the sunlight spectrum between the shortest visible rays and the longest X-rays. Normally, incoming UV is largely blocked by a thin but essential layer of ozone 10 to 20 miles above the earth, the altitude at which SSTs would fly. Some scientists fear that much of this ozone--a highly reactive molecule made of three oxygen atoms--would combine with the water and oxides of nitrogen in SST exhaust, thereby reducing the UV-blocking power of the ozone layer.

This layer is something we cannot afford to lose. The protective efficiency of ozone is so great that even a 5 per cent reduction over the mid-United States would cause a 26 per cent increase in UV radiation at harmful wavelengths; a 50 per cent decrease in ozone would yield a 1,000 per cent increase in radiation.

The effects of high doses of UV upon living things are so damaging that formation of the ozone layer was probably prerequisite to the evolution of life. Induction of the three most common malignant human skin tumors, for example, is correlated with the exposure to sunlight. A 5 per cent decline in ozone, says the panel, would

Copyright © 1973 by Saturday Review Co. Used with permission.

produce at least 8,000 "extra" cases of skin cancer per year among the U.S. white population (118,000 cases now are detected annually). In addition, UV causes skin changes that we normally blame on aging: wrinkling, discoloration, thinning, dryness, and blook vessel dilation. "Sunlight," continues the report, "is far more important than the passage of time in destroying the visage of youth."

UV also injures DNA, the essential carrier of genetic information in every cell. Although DNA has a limited capacity for self-repair, increased UV could overload the repair mechanisms, leading to deterioration and death of the cell or mutations among offspring.

The penetration ability of UV is especially worrisome. It passes through water easily enough to cause sunburn in bathers on cloudy days. The panel fears that an increase could upset ocean food chains by killing near-surface plankton. UV even can penetrate skin, muscle, and bone, reaching the depths of the brain. Some 25 to 75 per cent of UV radiation diffuses through woven fabrics, particularly the newer synthetics.

The theme of the report is smoothly summarized in this recommendation: "It must be made clear that life on earth is in a delicate balance between the beneficial and detrimental effects of sunlight." The very existence of the report, however, acknowledges the failure of the science community to supply data to the public during the SST information crisis. The panel recognizes this communications problem, chiding photochemists and photobiologists for not "demonstrating the relevance of their expertise to national problems."

ANOTHER POLLUTION DANGER:
TOO FEW IONS

Tired? Maybe your body hasn't been getting enough atmospheric ions lately. At least that's the theory of University of California bacteriologist Albert P. Krueger, one of the world's leading air ion researchers. Krueger has been experimenting with the effects of airborne ions on animals for almost two decades, publishing his findings in such journals as Proceedings of the Royal Society of Health (Britain) and, more recently, the International Journal of Biometeorology.

Mice, he observed, die more often from fungal, bacterial and viral diseases when the ratio of positive to negative ions is high in the air they breathe, or when the normal total number of ions falls abnormally low. Though Krueger has chosen to limit his own investigations to animals, he told Science News that pollution causes a marked decrease in atmospheric ions which, he believes, could affect human health. He said studies by colleagues in Japan and Israel show that human subjects become listless and irritable when ion concentration is decreased under laboratory conditions and that people who "feel" impending changes in the weather may be reacting to ion level changes. The mechanism of reaction remains unknown, Krueger said, but research indicates some connection between atmospheric ion levels and the level of a brain chemical called serotonin, which, among other things, affects blood pressure.

Reprinted by permission of Science News, © 1973 by Science Service.

MYSTERY DISEASE ARRIVES QUIETLY, STUNS WORKERS

Don Frederick, Jr.

Medical detectives hunt toxic culprit in plant's chemicals.

The print department at the Columbus Coated Fabrics Co. plant never has been a particularly pleasant place to work.

Employees are allowed two 10-minute breaks during their 8-hour shift for meals, smokes, and relaxation. So most workers just eat on the job.

The machines that are used to print designs on wall coverings, the company's main product, cause constant nicks and cuts on the hands. Many veteran employees can swap stories of arms trapped in the elaborate equipment. And then there are the ever-present fumes, caused by inks and solvents in the print machines.

Buddy Moore, a print-machine operator, calls the department "a glue-sniffer's paradise." But Corwin Smith, president of Local 487 of the Textile Workers Union of America, adds: "After a while you get used to the smell. You don't even notice the fumes."

LEG BRACES

Smith transferred from his print-department job in April. He left just in time. During this spring and summer, an illness began to afflict the print-shop workers. By mid-September almost 50 of the 130 or so of them and at least 10 others working in nearby areas were found to have "peripheral neuropathy," a nerve illness that weakens the body's limbs. By late September about 100 workers, including 2 at

Reprinted with permission from The National Observer, copyright Dow Jones & Company, Inc. 1973.

a plant a few miles away, showed some form of the disease.

Further testing continues to add to the numbers found to be stricken among the plant's 950 production employees. Medical authorities here believe the disease is caused by the inhalation, ingestion, and absorption of an industrial chemical.

Most of the ill workers have mild forms of the disease. But one worker now requires leg braces to walk. Five others have been hospitalized since the spring and may have suffered permanent crippling effects. About half the plant's workers decided to stay home last week. According to Smith, "They are afraid for their safety."

Dr. John Cashman, director of the Ohio Department of Health, terms the workers' illness a "tragic example" of a larger problem. "Myriads of new chemicals are introduced in thousands of factories each year," he says. Some of them cause "continued minor insults to the body" that can eventually result in sickness or take "10 years off a normal life span."

NEW PROCEDURES

Identifying the chemical at fault is "a medical detective story," according to Cashman. He says "the toxic agent could be any one of a number of chemicals or a particular combination of chemicals" and admits that gathering conclusive evidence may take months.

Meanwhile, the company, a division of Borden, Inc., has adopted new safety procedures. The print shop remains open, but most department employees, on a union suggestion, have not worked since Sept. 6. Many sit in the union hall and ponder why the danger signs went unrecognized.

It was last April that the first printshop employee became too weak to work. He was treated for arthritis. No one at the plant suspected anything unusual, certainly not the employee, James Osborne, 43, a maintenance worker for 18 years in the department. But Osborne knew that he was run-down. So in early May he quit a second job he had held for seven years. A few days later he was talking to some friends at the plant when his legs buckled. He thought it was the result of the heat and his run-down condition.

Osborne's condition worsened. His feet began to drag. They felt numb. By June he could not walk up a flight of stairs. He was hospitalized and treated for arthritis. When he returned to work in July, he heard that a few other print-shop men on various shifts were ill with symptoms similar to his. This seemed only a coincidence to him.

OUTWARDLY SERENE

Osborne grew weaker, and work became harder. In August his physician sent him to specialists in neurology. Late that month he mentioned to one the similar symptoms of his fellow workers. The specialist was suspicious. She made inquiries of the state health department. By most accounts, this was the first break in the case.

For Osborne, it may have come too late. He suffers from a severe form of the nerve disease. He has not worked since Aug. 20, and his condition has not improved. The slightest misstep will cause him to collapse. He says an easy chore like changing a tire "requires every ounce of strength I have." But he remains outwardly serene. "Trust in the Lord, and everything will work out," he says.

The state health department began searching for the common denominator in the lives of the six workers who had been stricken by that time. Dr. Donald Billmaier, chief of the division of occupational health, says that medical backgrounds and possible common use of a drug or alcohol were checked. But only print-department employment was common to the men. After a few weeks of testing, almost one-third of all print-shop workers, both male and female and of all ages, were found to suffer from varying forms of peripheral neuropathy.

Most of the employees were shocked. The clues had been present for months. Many workers had complained of weakness in the limbs, and a number had been limping during July and August. But few suspected their everyday environment was the cause of their fatigue. They blamed other conditions. Marian Handel, 37, has worked in the print department almost 16 years. To better his financial position, he worked overtime as much as possible. This summer he began to tire sooner each day. In his spare time he would just "lounge around and watch TV." He thought he was overworked; he stopped requesting overtime.

Vernon Johnson, 24, also worked overtime often. He needed the money to pay for a new home and swimming pool that he purchased soon after getting married in January. The extra work didn't bother him, but during the year he began to curtail other activities. He quit his bowling league. He stopped tinkering with a 1966 Impala Super Sport in which he had $7,000 invested. Then, at a state fair in August, he suddenly had trouble walking; according to his wife, Beverly, he was "limber." But Johnson was not concerned. Over the past two years he has been on an extended diet and has shed close to 100 pounds. He assumed his problems were due to the weight loss.

Charolette Goodrich, 27, is an attractive brunette who has worked near the print shop for about four months. She describes herself as "normally a very spunky person." She works the third shift, from 10:30 p.m. to 6:30 a.m. She used to do her housework after work, but in the late summer she began "falling in bed" instead. Charolette had suffered a concussion a few months ago; she blamed this for her lethargy.

SPECULATION BECOMES REALITY

Oakley Dingess, 38, began feeling weak in May. He thought he had a virus. But his difficulties increased. He couldn't lift equipment on the job. Soon, walking into the plant from the parking lot became a challenge for him. Finally, it was an effort to turn the ignition key in his car. Dingess spent much of the summer in the hospital. He knew of Osborne's similar condition, but heard it was treated as arthritis. There was no reason to suspect the print-shop chemicals. Dingess now suffers from a severe case of peripheral neuropathy. After spending a day sitting in the union hall, he will return home "and feel like I've worked a full eight-hour day."

In late August some workers took note of the spreading illness. James Sorrell, 23, and his friends joked about a "mysterious disease." Around the same time, a collection was taken to financially aid the stricken workers. Buddy Moore, 22, remembers wondering what he would do if illness forced him away from job and pay. A few weeks later this idle speculation became a reality. Both Sorrell and Moore now have a mild form of the nerve disease.

Though most employees first learned of their dangerous working conditions in early September, company officials say they conducted their own investigation in July. Joseph Recchi, director of company employee relations, says that "at that time the plant's physician was not convinced the sickness was industrially related."

CHEMICAL SUSPECT

More than a month later the state health department quickly determined the nerve disease was indeed industrially related. The chemical at fault has been harder to identify. Methyl Butyl Ketone (MBK), a chemical that thins ink used for designs, was the original suspect. MBK was introduced at the Columbus plant about a year ago. It was the one variable in an otherwise constant situation. The company has now stopped using the chemical. Ironically, the chemical that MBK replaced was discarded because of its high pollution level.

But because cases of peripheral neuropathy have been discovered in workers outside the print-department area, MBK is no longer the prime suspect. Other chemicals used throughout the plant are now being examined.

While the toxic chemical is sought, Dr. Billmaier believes, the newly implemented safety procedures, such as a new ventilation system, a prohibition of eating on job sites, and the required use of respirators, will protect the worker from harmful exposure. Dr. Cashman believes that the plant was guilty of "sloppy housekeeping," but that now the factory "may be the safest place in Columbus."

Company employees are understandably leery. As evidence of the illness spreads through the plant, union officials are considering a suggestion that workers outside the print-shop area boycott their jobs. Says union-local president Smith "Our people don't want to be guinea pigs."

ON HALF PAY

Print-department employees, though, admit they are in a frustrating position. Says Jerry Brooks, 41: "I don't particularly want to return to that shop, it's not 100 per cent safe yet. But I don't want to throw my 22 years of seniority down the drain either." Also, the various types of compensation that the workers collect when off the job amount to about half their normal wages.

For most of the stricken workers, a decision about returning to their jobs is not of immediate importance. The only therapy for peripheral neuropathy is rest; at least a few months for those with mild nerve damage, perhaps as long as four years for more severe cases. If the mysterious toxic chemical has destroyed any nerves, a possibility for a few of the men, then disability in the limbs will be permanent.

While they rest, the men and women must deal with problems created by the illness. Vernon Johnson may have to give up his new home and swimming pool; the money from workmen's compensation will not meet the payments. He and his wife had planned to adopt a child. That dream is now postponed. He must also adjust to his weakened state. Recently, his 18-year-old cousin had to open a twist-off bottle top for him.

FEAR OF BOATING

Bill Moore, 40 will have to forgo the outdoor life. He doubts that he's strong enough to do much hunting this fall. He's also afraid to go boating in his cabin cruiser. "If she flips over, there's no way I could make it back to shore," he says.

But the major problems will be psychological. Oakley Dingess says it's "nerve-racking" just thinking about the nerve disease. And all the ill workers speak of the "helplessness" of their condition, suffering from a disease of unknown origin and length that no medication can cure.

Dr. Cashman hopes that the plight of these workers will prod other factories in Ohio to improve their safety techniques and monitor industrial chemicals more carefully. "An industrial state lives on an earthquake fault," he says. For the workers at the Columbus Coated Fabrics plant, that ground is very shaky at the moment.

*

POWER PLANTS AND COTTONTAILS
Bruce Wallace

> Something must certainly be done about environmental decay, but the solutions seem to be even more ellusive than the problems.

An Environmental Impact Report was recently left on my desk. Idly I thumbed through its 200 or so mimeographed pages--maps, charts, tables, diagrams, and much officialese. The report had been prepared in fulfillment of federal regulations governing the construction of nuclear power stations. As I flipped the pages, I saw diagrams of water discharge nozzles; my eye picked out such words as "cottontail rabbit," "alewife," and "chipmunk." Curious assertions surfaced from the text: The danger of contamination of milk by Iodine-131 is not sufficient to worry about, but in any case there are no dairy herds near the site of the proposed plant.

Flipping pages is, with me, a reverse process, because I start from the end. At last I approached the front matter--Foreword, Table of Contents, and Summarized Cost-Benefit Analysis. I paused to read the summary because of my unfamiliarity with these studies. As I now recall, the summary analysis consisted of three "benefits" and four "costs." Benefits: 1) the production of 3,000 megawatts of electrical power; 2) the injection of $800,000 into annual regional payrolls; 3) an annual contribution of $4 million to local property taxes. Costs: 1) 45 acres of currently unused land for plant construction; 2) usurpation of 100 acres for additional transmission lines; 3) the dumping of some 700 pounds of salts per day into Lake Ontario; 4) the exposure of local populations to a level of radiation equal to 7 per cent of that which they now receive from natural sources.

Copyright © 1973 by Saturday Review Co. First appeared in *Saturday Review*, May 1973. Used with permission.

That was it. I reread the summary once more, searching in vain for something profound, for a shred of evidence that the report's authors were emotionally moved by, or that they stood in awe of, the task that had been assigned to them. None was to be found.

Now, 3,000 megawatts of electrical power is about one-fifth of the generating capacity of the entire United States in 1930, or one-tenth the capacity in 1940. In the face of such magnitude what does the cost-benefit analysis consist of? Local salaries and property taxes on the one hand, 700 pounds of salt per day on the other. Inside the report, not summarized, are the cottontail rabbits, the chipmunks, and the alewives. Oh, yes, and an explicit surmise that persons will continue in the future to demand television sets, electrical appliances, and air conditioners.

In my estimation, a report of the sort left on my desk is incredibly inadequate. In a sense, I expect reverence from a group of human beings who are assessing a power plant that is the equivalent of one-tenth of 1940 America. I expect them to look, or appear to be looking, one-tenth of the way to Los Angeles in a search for possible impacts. On the other hand, I am humble in knowing that those who compiled the report are fallible human beings making a living by doing the best possible within the confines permitted by their superiors. Would I have done better in their position? Not likely. The urban sociologist Jane Jacobs has described her frustration in working with zoning officials; because she cannot accept their basic premises, she is incapable of offering them sane advice. And I make this confession now: I would be a useless member of an environmental impact study group.

Can anyone seriously believe that the generation of 3,000 megawatts of electricity will have an environmental impact that is limited to less than 200 acres of land? Suppose that I am asked to prepare an impact statement concerning the use of my automobile. I shall anticipate the effects of starting it, pressing down the gas pedal, and throwing it into gear so that it rushes recklessly from my driveway at top speed. In preparing my report, am I to count and describe the gravel stones in my drive, to identify the stunted weeds that grow here and there among them, to study the numbers and habits of the ants and beetles that scurry back and forth in search of Lord knows what? Am I to prepare a report that claims on the cost side that three pounds of gravel will be distributed as the rear wheels spin? That several hundred small pebbles may be irretrievably lost when thrown into the neighbor's lawn? That there will be some destruction of the wildlife inhabiting my driveway but that the number of casualities involved should be small? That the populations are expected to recover quickly? Do I speak of the children at play in the park across the street? Of the residential area beyond the park? Of the shattered lives of those persons my car may strike if it runs amok?

It seems to me that the creation of 3,000 megawatts of electrical power has consequences that follow just as surely as those that can be expected in the wake of a careening car. I think these consequences belong in an impact report--not in the form of glib generalities about TV sets and electrical ovens but, rather, in specifics dealing with acreages for new factories and storage yards; the demands of these factories for raw materials; the tons of wastes that will be produced; the trucks needed to haul materials to the factories, finished products to the homes, and industrial waste and household garbage to the dumps; the highways on which the trucks will move; and all the other items that follow as surely as the seasons. To speak only of chipmunks and cottontails at the plant site and the alewives offshore is a travesty.

Throughout the report the "first straw" view prevailed. If a report is to enlighten its reader, it should be prepared with the opposite view, the "last straw" view. Take, for example, the 700 pounds of salt that are to be dumped each day into Lake Ontario. The lake holds some 400 cubic miles of water that already contain low concentrations of salt; consequently, the report argues, the planned addition of 700 pounds daily is trivial. Suppose, on the contrary, that the salt concentration in lake Ontario was exceedingly high already. Would this supposition alter the argument concerning the dumping of salt in the lake? Not at all! If there are already millions of pounds of salt per cubic mile in the lake, then the addition of a mere 700 pounds per day is also trivial. I submit that if a report can reach the same conclusion from two totally different sets of data, it is either useless or misleading.

To be of use in arriving at rational decisions, an environmental impact study, even though it is prepared for a single power plant, cannot pretend that its subject exists in a vacuum. On the contrary, it must describe the entire nationwide community of power plants, of which its subject is but one. It is not enough to be told that the one plant will discharge 200,000 curies of radioactive gas yearly into the atmosphere. To evaluate the meaning of such an admission requires that we know how many other plants are doing the same and where these other plants are located. If present power policies are to continue, how many such plants are anticipated in the future? Nothing of value is gained if the impact report for the 2,000th nuclear power plant emphasizes that the expected emissions of salt, heat, and radioactive wastes will amount to only five ten-thousandths of what is already in the atmosphere and oceans.

The individual environmental impact study (though it may be of some local value) cannot possibly forestall disastrous environmental changes. Indeed, if one were asked to design a procedure that was guaranteed to produce disasters, the individual impact report would be an obvious suggestion. By being constrained to an analysis of individual stations, one by one, none of the study groups will notice when the heat, salt, radioactive wastes, and the cascading ramifications of energy production have merged throughout the length and breadth of the land. And for those who prepare impact studies as well as for those who merely flip through them, there will then be no place to hide.

*

MOSQUITOES FACE DEADLY FOE

        Although the ultimate solutions to our environmental crisis will not be technological (see "The Historical Roots of Our Ecologic Crisis", and "The Tragedy of the Commons"), scientists have devised incredibly clever ways to deal with some specific problems. Most notable are recent developments in pest control. The following articles tell of some effective (if diabolical) new methods of insect extermination.

    Big, buzzing mosquitoes are a painful fact of life in the San Joaquin Valley.
    Because of the vast amount of irrigation and inevitable waste of water, billions of mosquitoes can be produced in a few days in one carelessly flooded pasture.
    These mosquitoes (aedis nigromaculis), while not disease carriers, are severe biters and can cause weight loss in livestock and secondary infections in humans and animals scratching the bites.
    They become a real uisance when they move into urban areas, but there is hope a revolutionary type of insecticide may help alleviate the problem.
    The compound is a synthetic copy of the juvenile hormone (JH) naturally produced in growing mosquitoes. A JH insecticide trade named Altosid has been under testing in the San Joaquin Valley for the last three years.
    Dr. Charles Schaefer, University of California entomologist and head of the UC Mosquito Control Research Laboratory here, thinks Altosid may prove an immensely effective tool for mosquito abatement districts in the valley.
    The traditional method in insecticide research has been to find toxic materials

Reprinted with permission of United Press International.

which would kill a particular insect faster than surrounding life forms, but hormones have presented researchers with exciting new, ecologically sound alternatives.

Scientists who first isolated insect hormones in the late 1960s found that massive doses of natural or artificial insect hormones applied to insects would throw their biochemical systems out of kilter. But surrounding life forms with different hormonal setups remained unaffected.

Altosid is the first synthetic hormone insecticide in history to go from research to testing to preliminary approval for limited public use by the Environmental Protection Agency (EPA).

Mosquitoes pass from egg to larvae to pupa to free flying adult in five days, and in order for Altosid to work, it must reach the insects in the middle of this process.

Dr. Schaefer explains how it works:

"At the time the larvae is ready to transform into the pupal stage, if we introduce a large amount of juvenile hormone, or a closely related compound that would cause the same type of activity, the normal transformation of a larvae into a pupa doesn't occur. The larvae transforms into an abnormal pupa and that pupa then dies."

Mosquitoes caught in the pupal stage also will be transformed into abnormal adults and die, but JH compounds are so specific they have no effect on adult mosquitoes.

Schaefer and his associates have been testing Altosid for three years and hope to get full EPA approval and registration for it by next year.

It has proved more than 95 per cent effective on larval and pupal mosquitoes but Schaefer hopes mosquito abatement districts will not immediately adopt its use on a wide scale.

"We feel it will be a useful tool," he says. "However, we are not advocating that mosquito abatement districts turn to this more expensive material and convert their programs to using it for two reasons.

"One reason is cost and the second is that we know that if we were to use it indiscriminately, if we were to have a low-cost compound, we would find resistance to it within a few years.

"There's no question that with any new type of chemical agent, if we use it in enough frequency, on a large enough population such as we have with mosquitoes, you are going to select our organisms that would be resistant to it."

Because of the high-development costs and the limited use for insect hormones, Altosid currently costs $75 per pound, although Schaefer has found the material will work at doses as low as 1-50th of a pound per acre.

Schaefer says the chief problem encountered so far is not with the insecticide but with humans.

"Our chief problem is mismanagement of land and water," he states. "In other words, if a person just keeps irrigating a pasture and grows grass for 10 or 20 inches deep over the top of the water, there's no possible way you can treat that water because the spray (applied by airplane) will be caught by the vegetation. With this kind of a field, there's nothing we can do about it."

Schaefer says the way to solve the mosquito problem is to eliminate places where mosquitoes breed and he says abatement districts are beginning to tighten regulations against wasting water and letting water accumulate in stagnant pools where mosquitoes can breed.

"That is the first and most important thing that can be done in the way of mosquito control," he emphasizes. "And then in chemical research we can provide tools that can actually allow suppresion of mosquitoes in areas where breeding places haven't been eliminated or where ditches break and suddenly there is flooding that wasn't expected."

INFECTIOUS CURE
Kevin P. Shea

On December 1, the United States Department of Agriculture (USDA) approved the use of an insect-killing virus to control the cotton bollworm, one of the most damaging pests of cotton. Although the registration was granted for only one year, in this age of pest resistance to synthetic chemicals and of increasing evidence of ecological damage from the use of chlorinated-hydrocarbon pesticides, like DDT, the announcement is considered a giant step in insect control.

Since the early 1940s, the USDA has registered perhaps 50,000 synthetic chemical formulations (see <u>Environment</u>, October 1970), but this is the first virus and only the third disease agent to receive federal approval. The other two, "milky disease" of the Japanese beetle and <u>Bacillus thuringiensis</u>, a disease which affects a variety of caterpillars, are both bacteria and have been used in the field with considerable success.

Insect viruses were recognized as far back as the Civil War, but only recently have they attracted attention as a practical means of controlling insects.

For a variety of economic and biologic reasons viruses have had a difficult time gaining federal sanction. Most important, probably, is the fact that federal officials were entirely without protocol as to how to proceed with the registration process. That is, they did not know what safety tests to require to assure that man and other animals would not be harmed should viruses come into wide use. In general, although the procedures have been similar to those for registering a synthetic chemical, both the Pesticide Regulation Division of the USDA and the Food and Drug Administration have seemed much more cautious.

Two biological facts have also slowed interest in the commercial development of

---

Reprinted by permission of The Committee for Environmental Information, Inc. Copyright © 1971 by The Committee for Environmental Information, Inc.

insect viruses and have thus prevented a rush to obtain federal registration. First, as a general rule, each virus is infective in relatively few species of insects, some in only one. This of course strictly limits the sales of such a product in any one year and thus reduces the return of capital invested in developing, registering, and marketing the product. This investment can run as high as several million dollars in the case of a synthetic chemical which has the advantage of being useful for a wide variety of pests. Secondly, viruses can only be produced in living cells, which makes mass production all the more difficult. At the present time the only practical way to produce massive quantities of virus material is to grow the host insect in large numbers, and artificially infect them, a process which is still relatively new. Recently, however, techniques have been devised which substitute artificial diets for the host insect's normal plant food and this has greatly facilitated production.

Other problems may arise in the production of viruses for insect control. Unlike the production of a chemical, the production of a biological material cannot be precisely controlled. During the early years of the mass production of <u>Bacillus thuringiensis</u> by the fermentation process, there were great differences between the formulations of the several companies that produced a commercially available product. One company guaranteed control of an insect by using twenty pounds of material per acre, while another company claimed control could be achieved by using only ten pounds per acre of its product. Both were probably correct, and the difference was actually a result of a difference in the fermentation process used or perhaps the strain of bacteria used in starting the fermentation process.

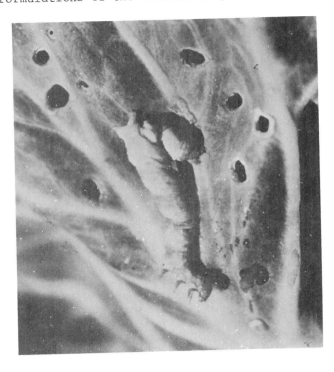

Dead larvae of the cabbage-looper, like the one shown here, have been collected, ground up, strained, and the virus material used successfully to control this pest of lettuce and other crops.

The federal label issued for the sale of <u>Bacillus thuringiensis</u> did not take these differences into account, and, although the USDA is charged with assuring the efficacy of all registered compounds, it could not do so in this case.

Even though viruses must be grown in living tissues, the same problems may arise in mass production. Some strains of a single virus are more or less virulent than others. A continuous program of testing could easily standardize the material being prepared for field use, however.

FORESTALLING RESISTANCE

The situation is doubly complicated by the fact that both the virus and the host insect are subject to change. This brings up another difficulty that will almost surely arise if viruses are used in the same indiscriminate way that synthetic chemicals have been used for so long--the development of resistant strains of insects. It will be important to monitor the effectiveness of the virus in each area in which it is used in order to determine whether insect populations arise that are less sensitive to the virus.

Although there are hundreds of known insect viruses, the Department of Agriculture has expressed the opinion that each one will require a separate registration. Registrations may come more quickly now that a protocol has been established, but still the expense of supplying the data upon which the registration is based is great when compared to the market for a single virus material.

There are about three hundred recognized viruses in four major categories, but only a few have been studied as potential

A group of grasshoppers killed by one of three hundred known insect viruses.

agents for pest control. The most promising at present are those members of the group known as the nuclear polyhedrosis viruses. Unlike viruses of other animals, their presence in insects can be detected by examination of insect tissue with an ordinary light microscope. In an infected insect, polyhedral (many-sided) shaped bodies can be seen in the cell nucleus. The polyhedra are bodies composed of protein material within which 10 to 100 submicroscopic virus particles are packed. As infection progresses in an individual insect, more polyhedral bodies are formed until the entire body of the insect is a mass of encapsulated virus particles. The skin of the dead insect then ruptures and spills its contents into the environment. When another insect eats plant material upon which some of the material from a virus-killed insect has spilled, the polyhedra are taken into the gut, where they are dissolved. This releases the virus particles which then invade the cells of the new host, replicate, and produce more polyhedral bodies.

A second group of insect viruses, the granuloses viruses (viruses that appear as a mass of granular material in cells), produces a capsular form of proteinaceous covering smaller in size than the polyhedral viruses and can invade either the cytoplasm or both the cytoplasm and the nucleus of the cell. A third group, the polymorphic (many-shaped) viruses, produce capsules of a variety of shapes and only invade the cytoplasm. The fourth group, the noninclusion viruses (those that do not occur in bundles with a protein cover), do not produce any kind of capsules and are present only as virus particles.

Because of the wide variety and distribution of virus materials, they can be found in almost all kinds of insect habitats. This does not, however, guarantee that an insect population will be subjected to an outbreak of a viral disease. Because of the low mobility of viruses, most outbreaks occur only when the population reaches a very high density and members of the population come into close contact. In times of crowding, viruses which may be present in the population can become virulent due to stress from overcrowding and shortage of food. If this happens in a cultivated crop, a great deal of damage can be done before the virus infection checks the insect population. However, if enough virus material can be uniformly applied to the crop, pests can be controlled at low densities before any appreciable damage is done. The object then is to get the material in the right place at the right time in the proper concentration.

In spite of the fact that official recognition of the usefulness of viruses has been ponderously slow, some pest control experts have been using "bootleg" viruses for some time. But taking advantage of naturally occurring outbreaks, one pest control consultant in Califronia, in a single season, supplied enough material to growers to treat 25,000 acres of lettuce and cole crops (cauliflower, broccoli, cabbage) for control of the cabbage looper. He simply gathered the infected larvae which were then pulverized, strained, and frozen until a potential outbreak of cabbage looper was discovered. The "soup" was then thawed, mixed with water and sprayed on the fields at the rate of seven to ten larval equivalents per acre (one larval equivalent, or the amount of polyhedral bodies produced by a full-grown larva, contains about one hundred million polyhedral bodies). His charge to clients was one dollar per acre for just the material, a price that is highly competitive with synthetic organic chemicals. Although he has been using this kind of control for ten years, he was recently ordered by the California Department of Agriculture to discontinue the practice because the looper virus is not a registered chemical in that state.

On the other hand, the states of South Carolina and Tennessee have both given approval for use of the cabbage looper virus on an experimental basis with no limitation on the amount of acreage that can be treated. Growers in those two states have been plagued by looper populations that are highly resistant to the chemicals approved for use on lettuce and cole crops and a new method of control was urgently needed.

Virus preparations have also been sold on a limited scale in Texas for cotton bollworm control, on the Atlantic seaboard states and in Arizona for the control of cabbage looper, and in Indiana for the control of the pine sawfly.

Should the use of viruses prove successful and safe on a wide scale, the advantages over synthetic chemicals would be numerous. In fact, the characteristics of viruses are just the opposite of those characteristics which have caused ecological problems from the use of synthetics. They are not persistent; they are not toxic to non-target organisms; and they can be used safely by growers. They are naturally occurring materials that are simply manipulated to put them to their most advantageous use.

Viruses and bacteria are not the only candidates for use in insect control. Fungal and protozoan diseases might also be employed. There are over one-thousand recorded diseases of insects from which to choose and it appears that an entirely new approach to insect control is beginning to gain impetus.

AUSSIES TURN TO BEETLES IN ECOLOGY FIGHT
David Lamb

Insects aren't the only problem that agriculturists have to deal with. Cattlemen in Australia may have found an ecologically sound solution to their unique problem.

Australia is importing thousands of dung beetles from Africa to combat one of the major economic hazards facing the ranching industry--an abundance of cow manure.

The 15-year project, being watched by biologists throughout the world, is designed to introduce to Australia aspecies of tunneling dung beetles that can bury the manure now smothering the land.

Although native beetles disperse cattle manure in other parts of the world, nature did not put them in Australia because cattle arrived here only 150 years ago after being imported from Asia.

Australia now has 30 million head of cattle and each accounts for about 10 "pats" a day. At any given time, 3 million acres are covered by the pats, which harden and lay undisturbed for months or even years and thus kill the growth needed for grazing.

But in experiments already termed successful in parts of Australia, the tiny onthophagus gazella beetles tunnel up beneath the pats, separate them into pinhead-sized balls and roll them into their shafts. Within 48 hours, the pats have disappeared.

In addition to clearing the land, this enriches the soil because the nitrogen in fresh dung is lost to the atmosphere when the manure dries on the surface and

---

Copyright 1973 by Los Angeles Times. Reprinted by permission.

it kills the eggs of parasitic worms in the pats.

"The rate of the pats' disappearance is really quite spectacular," said Dr. Douglas Waterhouse of the Australian Commonwealth Scientific and Industrial Research Organization (CSIRO).

Waterhouse said, however, that more than 200 species from south of the Sahara would have to be imported, tested and bred. Twenty species have been imported since the program's inception in 1967.

The beetles are dispersed at selected sites and allowed to spread through natural migration. They have already colonized 250 square miles in northern Queensland and penetrated 50 miles inland.

Dr. G. F. Bornemisszsa of CSIRO's division of entomology heads a 10-man Australian research team in Africa and describes the project as "probably unique in the history of applied ecology."

One of the important benefits of the beetles' work also is the elimination of the bush fly and the buffalo fly which breed in fresh manure.

The pesty bush flies are so abundant in some areas that the growth of tourism has been impeded and children wear veil-like insect screens over their heads.

And the blood-sucking buffalo fly affects meat production by irritating the cattle so badly they cannot eat. Pesticides have not eased the problem because, although they kill the insect they leave a residue in the meat which is unacceptable to foreign markets.

There are 250 indigenous dung beetles in Australia which bury the pellet-sized droppings of kangaroos and other marsupials but they cannot cope with the large pats of the domestic animals introduced by Europeans.

The CSIRO also has been responsible for two other world-celebrated achievements in biological control--a dramatic reduction in the number of rabbits, which compete with sheep for food and water, and destruction of the prickly pear.

In 1950, the first year the rapidly spreading myzomatosis virus was proven an effective control, hundreds of millions of rabbits died. The value of the wool clip increased $70 million that year.

The prickly pear, which came to Australia from America, had spread over 65 million acres of grazing land in Queensland and new South Wales by 1925. Stock literally was pushed from the land and abandoned homesteads often were invaded by the pear. The Queensland government then estimated it would cost $200 million to clear the land.

Thirteen species of cacti-dwelling caterpillars were brought to Australia from North and South America by the commonwealth prickly pear board. The coctoblastis caterpillar, which tunnels into the cactus, was one of them and within a few years it had caused the collapse of the vast pear jungle.

The Queenslanders of Dollby honored the caterpillar with a stone monument which still stands in the town square to this day. And the residents of Boorgara named their civil building the memorial hall to cactoblastis.

Although the ultimate solutions to our environmental problems will not be technological (see "The Historical Roots of Our Ecologic Crisis" and "The Tragedy of the Commons"), scientists have devised incredibly clever ways to deal with some specific problems. Most notable are recent developments in pest control. The next two articles tell of some effective (if diabolical) new methods of insect extermination.

*

THE CASE OF THE CHINESE CLAMS: WHAT TO DO?

Australian ranchers have high hopes for their new dung beetles, but sometimes foreign species can bring nothing but trouble to a new environment. The burgeoning Chinese clam population in North America is a dramatic example of Malthusiam population growth gone wild (see "An Essay on the Principle of Population").

They were first spotted in Oregon in 1938, after having gone undetected for possibly as much as half a century. Two years later they were found in San Francisco. Slowly they began moving eastward--it was 1956 before they were reported in Colorado--but then picked up speed. The following year they showed up in Ohio and moved down the Mississippi to Louisiana by 1962, thence spreading east to Florida by 1965, westward to Texas by 1968 and up the East Coast to Georgia by 1971.

Now they're up to the Delaware River, between Philadelphia and Trenton, Corbicula manilensis is here to stay.

C. manilensis is the Chinese clam, or perhaps more properly the oriental clam, since it exists throughout Asia and, in fact, was first found in the Philippines. Its foothold in the United States, however, is, to put it mildly, secure.

"It doesn't seem to have any natural enemies in this country," says Samuel L. H. Fuller, an invertebrate zoologist at the Academy of Natural Sciences of Philadelphia, who also found the creature last year in the Pee Dee River in South

Reprinted by permission of *Science News*, © 1973 by Science Service.

Carolina. Thus it is capable of getting a firm foothold in areas where the native species that must compete with it are developing more slowly.

In 1952, communities of the clams were found living on the bottom of California's Delta-Mendota Canal, which had been open little more than a year. They seemed to pose little if any reason for concern until the canal was partially drained in 1969. Engineers found the prolific creatures lining the canal in layers three feet thick, with as many as 5,000 clams nestled into a single cubic foot. It took a month and a half of shoving with bulldozers to clear out the 50,000 cubic yards of clams that had infested the canal bottom.

Nonetheless, both Fuller and R. Tucker Abbott, a conchologist with the Delaware Museum of Natural History, agree that more study is needed before the clam is branded a major ecological hazard. Even then, Abbott believes, preventive or even curative approaches are unlikely to have much effect. "There's no sense in spending a nickel trying to prevent it from spreading," he says. "I don't think it can be stopped. It's like trying to bail out Boston Harbor with a teacup."

How the clams have spread across the country is unknown. Fishermen using them as bait may help inadvertently by dumping buckets of excess clams in previously unexposed rivers. Bottom material dumped from one river into another by construction engineers could be another factor, as could shell collectors or scientists disposing of excess samples.

An important experiment that ought to be done in the near future, Abbott maintains, would be to feed the clams to ducks to see if they come out alive in the ducks' feces. The shells may well be sufficient to protect <u>Corbicula</u> from digestive juices and high temperatures, he says, and the ducks would then suggest an efficient potential long-distance transportation system.

The clams still have places to go. New York and at least parts of New England are apparently not too far north, since the creatures are known in Oregon's Columbia River, and Abbott foresees them as far south as Panama.

Fortunately, they're at least edible. In Florida on a shell-collecting trip last month, Abbott says, "I made a great New England clam chowder. But they needed a little salt."

MAN AND NATURE: SYMBIOSIS

> We should appreciate our beauty, as well as abhor our ugliness.

The realization that man, in his struggle to achieve civilization, has marred the earth's natural beauty and threatened its health, has made mankind seem, in some eyes, like a pestilence or parasite that sucks earth dry of its resources and upsets the balance of nature.

This view, according to Rene J. Dubos, is wrong: Man's relationship to the earth is not parasitic but symbiotic. Lecturing last week at the Washington meeting of the American Association for the Advancement of Science, the noted humanist elaborated on his view that man can, and does, improve on nature. "Nature is incapable, by itself, of fully expressing the diversified potentialities of the earth." Before the coming of man, he declared, earth was covered with forests and marshes. "There was grandeur in this seemingly endless green mantle, but it was a monotonous grandeur ..." Man, by clearing fields, erecting graceful buildings, planting gardens and parks--in short, by "humanizing" the earth--revealed the underlying diversity of the earth. "The symbiotic interplay between man and nature has often generated ecosystems more diversified and interesting than those occurring in the state of wilderness."

When it comes to solving ecological problems, he said, nature does not always know best. The periodic population crashes of lemmings, muskrats and rabbits are a clumsy way of reestablishing an equilibrium between population size and natural resources. As for the recycling processes considered to be the earmarks of

Reprinted by permission of *Science News*, © 1973 by Science Service.

ecological equilibrium, the accumulation of coal and peat demonstrates that nature has failed again. In fact, Dubos pointed out that man has completed the circle by burning peat, coal and oil, making carbon and minerals once more available for plant growth. The problems arise because man is recycling too rapidly, overloading the system. Nature has "junkyards," too: "The science of paleontology is built on them."

Dubos acknowledges that "many of man's interventions into nature have, of course, been catastrophic." But he believes that with wise management "mankind can act as steward of the earth for the sake of the future."

PART IV
MOLECULAR BIOLOGY

VEGETARIANISM: FAD, FAITH, OR FACT?
Sanat K. Majumder

> As consciousness of ourselves as complex chemical systems has developed, we have become increasingly concerned about the nutritional value of the food we eat. Sanat Majumber looks at today's most widespread nutritional movement, vegetarianism, from dietary and ecological points of view.

In the context of our enhanced interest in the mechanisms of human behavior and associated research, the aphorism "You are what you eat" appears highly simplistic and controversial. The fact remains, however, that little effort has been made to determine the relationship between types of food on the one hand, and the mind-body complex on the other, except with regard to administration of a specific medicine or the effects of a specific drug. I wish to deal here with one aspect of this question--vegetarianism--and review its relevance from historical, biological, and contemporary perspectives.

HISTORICAL OBSERVATIONS

Vegetarianism has never been without its advocates during any period of world history (1). Rather than a well-defined credo, it constitutes an individualistic regime of diet, and while a strict vegetarian excludes from his diet not only meat and fish but also eggs, there are some egg-eating "vegetarians" as well. In certain cultures this food habit has often been determined by the prevailing attitudes

From *The American Scientist*, March-April 1972. Reprinted by permission.

toward nature, religious beliefs, or, simply, nutrition. Needless to say, scientific observations with regard to vegetarian diet did not begin to appear until the turn of the twentieth century.

To most of us, food is a matter of taste and habit, occasionally predicated upon such items of controversy as cholesterol, calorie, sugar, and salt content. To many vegetarians, however, food is of the utmost importance in leading a contented, harmonious life. Contrary to popular belief, objection to animal slaughter and consumption of flesh per se constitutes only an insignificant part of vegetarian motivation.

One opinion holds that a well-balanced vegetarian diet encourages development of the intellect, increases the capacity for mental labor, and promotes longevity. "Intemperance which is the chief cause of pauperism and crime may be greatly discouraged by cultivation of vegetarianism" (2). Sir Henry Thompson affirmed that vegetarianism simplified human character and enabled the mind to enjoy more rest and, perhaps, more acuteness (3). Gandhi, the great Indian leader, made many empirical observations on dietary habits in relation to personality development. He tried various types of food and finally chose the simplest ones (fresh fruits and nuts were his favorite) for "calming of spirit and allaying animal passion" (4).

Religious arguments opposing the intake of animal flesh are quite numerous. Monks often abstain from meat, considering it to be a luxury and, therefore, at variance with their motto of simple living. Respect for animal life is the principal reason for vegetarian diet among the Buddhists (5). Certain sects of Hindus use the same argument for their dietary habit. Christian objection to meat can be lifted directly from the Bible in such quotations as "Meat commendeth us not to God," "I will eat no flesh while the world standeth," and "Be not among riotous eaters of flesh" (6). Two contemporary Protestant sects--the Bible Christians and the Seventh Day Adventists--require of their followers a vegetarian food habit.

The year 1887 was the jubilee year of vegetarian propaganda in England. In conjunction with this, A. F. Hill published a collection of essays, one of which aptly revealed his feelings: "Love underlies vegetarianism--in love, vegetarianism holds communion with God whose nature and whose name is love" (7). Anna Kingsford, a noted vegetarian of the nineteenth century and a medical practitioner, devoted much of her writing to the question of man's basic nature--herbivorous, carnivorous, or omnivorous. Her theory sought support from scientific data on comparative anatomy and pointed to the fact that the strongest animals--horses, elephants, and camels--were not carnivorous. She also referred to the amazing ability of athletes in ancient Greece who followed vegetarian diets (8).

Probably the greatest American endeavor in vegetarianism was the establishment of the Battle Creek Sanatorium in the late nineteenth century. The primary purpose of this institution was to provide medical care through diet. Significant in this connection was Dr. J. H. Kellogg's research and eventual success in the development of breakfast cereals.

Finally, we come to the emotional "back to the land" appeal among many of today's young people. This has imparted a new meaning to vegetarian habits that are being experimented with and sought by members of the "Woodstock generation" (9). The status of vegetarianism appears to be a natural outgrowth of problem-ridden industrial societies, causing many young people to look for a new economy that does not encourage abuse of technology, pure produce that does not require the use of chemical fertilizers or pesticides, and possibly altered spiritual values that sustain life.

BIOLOGY OF VEGETARIANISM

The scientifically oriented modern generation may look upon the foregoing historical observations with some degree of skepticism and limit the question of vegetarianism to certain places and people with peculiar cultural and intellectual characteristics. If vegetarianism is to rise completely above the alleged status of a fad or blind faith, it must satisfy three major criteria: (1) from the nutritional point of view, vegetable diets must be complete with the standard daily requirements of the human body; (2) they must alleviate, if not eliminate, the food crisis in certain parts of the world; and (3) they must make positive contributions to the enhancement of "bioethics" among the members of the next generation with regard to wild life, agriculture, and conservation of natural resources.

Biologically, it is well known that the efficiency of energy transfer from one trophic level to another is substantially lower than 100 percent. In other words, only a small fraction of the sun's energy that is transformed into chemical energy by green plants (producers) is actually passed on to the herbivores that consume plants as food. Various carnivores that consume herbivores receive, in turn, decreasing shares of the total energy budget. To use a familiar example, a tuna fish, for each pound of canned product, must have consumed 10 pounds of herring, which, in turn, required 100 pounds of prey that lived on, proportionately, 1,000 pounds of algae. The same energy distribution is applicable to ranch and forage crops that pay us dividends in the form of beefsteaks.

It is obvious that the harvest of available energy would be more efficient if more omnivores could be herbivorous. Many such animals (e.g. raccoons) change their food habit seasonally under the pressure of food scarcity, but we, the rational omnivores, can experiment with our food and respond to similar pressures with choices that are supported by cultural feasibility and nutritional studies. Scientifically, then, this argument may form the basis for vegetarianism.

The questions now arise: What are the nutritional requirements for good health? and Can vegetarian diets be balanced to meet these requirements? According to Clara Mae Taylor (10), a moderately active average man of 154 pounds and an average woman of 123 pounds need 3,000 calories and 2,500 calories, respectively, in daily food intake. In addition to appropriate amounts of carbohydrate and fat, the distribution of other essential components of our diet appears as follows: protein, 60-70 g; calcium, 0.8 g; iron, 12 mg; vitamin A, 5,000 IU (International Unit); thiamine (B_1), 1.5-1.8 mg; riboflavin (B_2), 2.3-2.7 mg; ascorbic acid (vitamin C), 70-75 mg.

Needless to say, this prescription is not undisputed. The iron requirement for women is said to be greater than that for men. The controversy over the efficacy of massive doses of vitamin C is resisting the common cold is another case in point. Linus Pauling (11) claims that a larger amount of vitamin C should be recommended for the human system. Many physicians, including Dr. Frederick Stare of Harvard University, on the other hand, question Pauling's assertion because the human body readily eliminates any excess of this vitamin.

In a vegetarian diet, the choice of a protein source does indeed pose some measure of difficulty. Animal flesh is ordinarily fortified with the ten essential amino acids human being need in their diet (12); a single vegetable source is rarely complete in this respect. Vegetable proteins are known as "second class" proteins owing to the absence of one or two essential amino acids. Besides, vegetable food is less concentrated owing to its very high percentage of water content, a fact that dictates consumption of a larger quantity for the same food value. Vegetarians, of necessity, must combine several food types to meet their daily need for protein. Such mixtures will inevitably include milk, cheese (lacto-albumin), soybean products (glycinin), Brazil nuts (excelsin), and enriched corn and wheat (glutenin). In any event, "a completely adequate protein supply can be attained solely from vegetable sources if the supply of amino acids is carefully looked after" (13). I shall return to this point below.

Obtaining an adequate amount of calcium and iron in the diet is a relatively easy task. If half the total daily calories are taken in the form of milk, cheese, fruits, and vegetables, a person will be fairly sure of a liberal calcium supply (14). Kidney and lima beans are especially high in iron content, and so are many leafy vegetables. Whole wheat bread and peanut butter are excellent supplements in this connection.

A vegetarian faces very little difficulty in finding an adequate vitamin supply. One-half cup of steamed spinach, for instance, provides over three times the daily vitamin A requirement. Turnips, dandelion greens, and broccoli are also known to be very rich in this vitamin. For thiamin (B_1), a vegetarian looks to such items as soybeans, lima beans, peas, wheat germ, and salad greens. Riboflavin (B_2), on the other hand, can be obtained from milk and vegetables such as soybeans, spinach, and asparagus.

Much research has been done on vitamin C (ascorbic acid) since the days when Captain Cook's men used citrus fruits to prevent scurvy. One medium orange or grapefruit supplies about 85 mg of vitamin C-- well over the daily need. Drinks prepared and processed from rose hips are known to contain substantially more vitamin C than citrus drinks, but they have yet to be introduced commercially.

ECONOMIC AND ECOLOGICAL IMPLICATIONS

A vast desert of starvation and malnutrition has spread ominously in today's overpopulated world. By comparison, the oases of affluence are very few and far between. The luxury of considering a balanced nutrition is indeed a cruel hoax in a society that exists at subsistence level. The fact remains, however, that with the highly improved communication and transportation systems of today, as well as our enhanced knowledge of various cultural traditions, the benefits of scientific discoveries can be made widely available to help alleviate the situation.

Of initial concern is the caloric requirement of people struggling to survive and the type of food to which they are accustomed. Until agriculture is diversified and expanded in the regions concerned, the major reliance for improving diet must be placed on cereal grains, which constitute the primary sources of both proteins and carbohydrates in almost two-thirds of the world. As indicated earlier, to raise livestock for meat, vast amounts of grains are needed as feed, and the resulting protein yield per acre is low (15). Calculations of the average pound-to-acre ratio reveal that plants have a greater efficiency for the production of eight of the essential amino acids. For instance, crops such as soybeans, peas, and beans yield 13 pounds of these amino acids per acre; others, including carrots, potatoes, cauliflower, brown rice, and cabbage, yield approximately 4.2 pounds per acre, whereas the corresponding figure is only 1.6 pounds for poultry, beef, lamb, and pork (16).

Coupled with the above observations, it should be noted that, at the suggested rate of one acre per capita, reclamation of new arable lands cannot begin to keep pace with the runaway world population, which is projected to be between six and seven billion by the year 2000 (17). Reclaimed lands in any case do not necessarily support the most desirable crops. Motivation for cultivation can be determined by "nutrition per acre" or "nutrition per dollar."

The challenge of feeding the millions is simultaneously a challenge to find new types of food and to engage in intensive research associated with human food habits. Some fundamental advances have been made in these respects by plant scientists. Corn, the staple for many people in South America and Africa, is, for example, naturally deficient in lysine and, to a lesser extent, in tryptophan. Mertz in 1963 discovered a high-lysine strain--Opaque-22--and subjected it to intensive tests in Colombia in 1969. Both Opaque-2 and Floury-2 (18) record increased lysin and tryptophan content to equal approximately 90 percent of the nutrition of skim milk. Under proper agronomic conditions, therefore, more than 200 million people who live in the tropical corn regions can derive dietary benefit, both quantitatively and qualitatively, from these strains.

Similar work with wheat has resulted in the development of varieties that are resistant to leaf rust and have greater protein content. Nobel Laureate Norman Borlaug developed a high-yielding dwarf Mexican wheat. His meticulous selection of genes and subsequent breeding program have opened the way to a wider distribution of this variety. Many tropical countries in South America and Asia are beginning to harvest dividends from this magnificent research. In a similar effort, by crossing a short Taiwan rice with an Indonesian variety, the International Rice Research Institute in the Philippines released in 1966 the much talked-about "miracle rice," IR-8. This new variety is dwarf, early maturing, highly responsive to nitrogen fertilizer, and very high yielding. Development of this variety could not have happened at a more crucial time or in a needier region of the world (19).

The unsung hero in the area of food research is the soybean, which was originally introduced as a livestock feed. Until the 1930s, the soybean was shrouded in an "oriental mystique." The production of soybeans in the United States has increased from 4.8 million bushels in 1925 to 840 million bushels in 1965. In view of its high protein (39-46%), phosphorus, iron, and calcium content, the soybean today is being processed for human consumption (e.g. soybean "bacon," "yogurt," and "cookies"). With ingenuity and appropriate promotion, this livestock feed of yesterday--500 million bushels of which is exported annually by the United States--can legitimately prove to be both an economic and nutritional bonanza for many people today (Table 1).

The problem of malnutrition may appear to be less dramatic than death from starvation, yet it is proving to be the greatest stumbling block during the formative period of millions of children. There seems to be little disagreement among scientists that a continuous protein-deficient diet produces irreversible damage to the brain. How can we resolve the basic question of filling the stomach and reducing malnutrition among people who cannot afford the taxing economy of livestock breeding for protein or who are traditionally or habitually vegetarian? The answer apparently lies in the extensive use of staples mixed with inexpensive but highly nutritive protein concentrates (20).

The ingenious approach of William D. Gray (21) is noteworthy in this connection. Mindful of the waste that occurs in the natural energy transfer, Professor Gray sought to "harvest" protein more inexpensively. He used waste substrate such as paper pulp, in which certain species of fungi grew profusely. In the presence of this "recycled" carbohydrate and added nitrogen fertilizer, the fungi readily synthesized protein. The dry, clear, tasteless, and odorless mycelium of the fungi could then be powdered and mixed with flour and other staples to augment the protein content from 2.5 to 15 percent.

"Although the carnivorous Americans are not likely to stampede for the protein-rich flour as a substitute for steaks, the prospect of its use in some developing countries cannot be overemphasized" (22). It is not as condescending for an affluent society to export protein-rich flour to people who are either vegetarians or whose dietary habit, of necessity, is cereal-based as it is to instruct those same people about developing a meat industry.

Table 1. Contributions of dried soybeans, other dried beans, and meats (average portions*) in percent of daily requirements.

Food	Calories	Protein	Calcium	Iron	Vitamins A	B_1	B_2
Yellow soybean	10	49.3	26	48.8	1.8	72	86
Kidney bean	10	25	17	72.5	---	24	11
Navy bean	10	28	16	72.5	---	24	11
Roast lamb	7.4	32.4	1.67	15.38	---	9.64	12
Roast veal	5.8	44.25	2.61	34.8	---	20	11

* An average portion of dried beans is 3 ounces, or about one cup of cooked beans; an average portion of meat equals 4 ounces. (From Williams-Heller and McCarthy, ref. 23.)

Flavored beverages enriched with protein are beginning to receive the attention of many commercial enterprises. Based on oil-seed meals, these beverages include Vitasoy of Hong Kong, Saci of Brazil, marketed by the Coca Cola Company, and the banana-flavored Puma, introduced by Monsanto. Among the mixtures of grains and oil-seed products, Incaparina (38% cotton-seed flour, 58% corn, yeast, and vitamin A) is probably the most popular example. In India, peanut flour and Bengal gram together constitute a multipurpose food. Sardiele is a soy and sesame mixture used by the Indonesians. Foods consisting of rice, wheat, soy, and peanuts have appeared in the Taiwan market. In Uganda, the common ingredients of similar food mixture are dry skim milk, sucrose, cottonseed oil, corn flour, and peanuts.

CONCLUSIONS

What I have discussed here neither advocates vegetarianism nor does it argue against such a dietary regime. Instead, I have attempted to examine with objectivity the question of vegetarian habits in a new perspective. The present dilemma of uneven economic growth in an overpopulated world necessarily forces us to reexamine basic concepts of human nutrition and to probe for simple, inexpensive, and nutritious food.

Although substantial evidence indicates that vegetarian diets can be balanced to meet human requirements, research in this area is very limited. The scientific communities of affluent nations have an obligation to initiate such research in view of the frightening statistics which indicate that two-thirds of all the children in the world--the future citizens of our planet--suffer from malnutrition owing to protein deficiency, while meat and fish proteins are out of their economic reach.

Even less is understood about the primary effects of different types of food on human personality and behavior. I am not talking about the secondary (social) effects, such as the apparent linearity of "starchy food--obesity--self-consciousness." Some empirical observations by early vegetarians, such as "calming of spirit," "allaying animal passion," and "acuteness of mind" associated with their diets, can and should be subjected to scientific scrutiny.

There is a vast difference between people who are vegetarians by choice and those who have no other alternative, economically speaking. Both groups can benefit, however, from the assurance that their diets can be made both plentiful and healthful. Fitting the mosaic of human cultures, the multiplicity of food habits and the availability of alternate foods should be welcome in this shrinking world. The control of a precipitous population explosion and the development of inexpensive but nutritious food must be matters of primary concern for all of us. While scientists can and must provide nutritional data as common denominators for the choice of food, they can hardly solve the human paradox of knowledge and wisdom, as indicated by the following Associated Press release on June 17, 1971:

> The diet of affluent Americans able to afford any food they choose is less nutritious than it was a decade ago. The problem: we are overfed but remain undernourished. As a result, experts say, 10 percent of the population may be anemic and 25 percent overweight.

REFERENCES

1. Alexander Bryce. 1912. _World Theories of Diet_. London: Longmans, Green and Co.
2. Charles O. Groom-Napier. 1875. _Vegetarianism, a Cure for Intemperance_. London: William Tweedie.
3. Sir Henry Thompson. 1898. "Why Vegetarian?" From _Science Paper No. 2_, The Egyptian Newspaper Press.
4. M. K. Gandhi. 1949. _Diet and Diet Reform_. Ahmedabad, India: Navajivan Publishing House.
5. E. W. Hopkins. 1906. The Buddhistic rule against eating meat. _J. Amer. Oriental Soc._ 27:455.
6. Gerald Carson. 1971. _Cornflake Crusade_. New York: Rinehart and Co.
7. A. F. Hill. 1897. _Vegetarian Essays_. London: Ideal Publishing Union.
8. Anna Kingsford. 1885. _The Perfect Way in Diet_. London: Kegan, Paul, Trench and Co.
9. _Time_. The Kosher of the Counterculture. November 16, 1970, pp. 59-63.
10. Clara Mae Taylor. 1942. _Food Values in Shares and Weights_. N.Y. Macmillan.
11. Linus Pauling. 1970. _Vitamin C and the Common Cold_. San Francisco: W. H. Freeman.
12. W. C. Rose. 1949. Amino Acid requirements of man. Federation of American Societies for Experimental Biology, _Proceedings_ 8:546-52. Note that Rose lists only eight amino acids as "essential."
13. Louis Bean. 1961. _Food and People_. U. S. Congress. Joint Economic Committee. Report of the Subcommittee on Foreign Economic Policy. Washington, D. C.: U. S. Govt. Print. Off.

14. Henry C. Sherman and Caroline S. Langford. 1943. *An Introduction to Foods and Nutrition*. N. Y.: Macmillan.
15. Ralph McCabe. 1961. *Food and People*. U. S. Congress Joint Economic Committee. Report of the Subcommittee on Foreign Economic Policy. Washington, D. C.: U. S. Govt. Print. Off.
16. M. G. Lambou et al. 1966. Cottonseed's role in a hungry world. *Economic Bot.* 20:256.
17. K. W. King. 1971. The place of vegetables in meeting the food needs in emerging nations. *Economic Bot.* 25:6.
18. E. T. Mertz, L. S. Bates, and O. E. Nelson. 1964. Mutant gene that changes protein composition and increases lysine content of maize endosperm. *Science* 145:279.
19. Lester Brown. 1970. *Seeds of Change: The Green Revolution and the Development in the 1970's*. N. Y.: Praeger.
20. Proceedings of the symposium on Integrated Research in Economic Botany VII: Protein for Food. 1968. *Economic Bot.* 22:3-50.
21. William D. Gray. 1965. *Activities Report 17*. Research and Development Associates.
22. Sanat K. Majumder. 1971. *The Drama of Man and Nature*. Columbus, Ohio: Charles E. Merrill, p. 104.
23. A. Williams-Heller and J. McCarthy. 1944. *Soybeans from Soup to Nuts*. N. Y.: The Vanguard Press, p. 12.

DIET FOR A SMALL PLANET
Frances Moore Lappe

If you have ever wondered why protein is so important, or what foods provide "good" protein, you should read this chapter from Frances Moore Lappe's beautiful book, <u>Diet for a Small Planet</u>.

Having read of the enormous food resources man will willingly squander in the production of meat, you might easily conclude that meat--and meat in large quantities--must be indispensable to human well-being and endowed with qualities unmatched by other foods. This isn't the case, as I hope to demonstrate in this chapter. Hopefully this discussion will be useful to anyone wishing to rely more on plant protein and less on meat protein for <u>any</u> reason--be it ecological, ethical, financial, or medical.

A. <u>WHO NEEDS PROTEIN ANYWAY?</u>

Why can't people get by on a diet consisting solely of fats and carbohydrates? In the first place, while carbohydrates, fats, and protein all provide carbon, hydrogen, and oxygen, <u>only protein</u> contains nitrogen, sulfur, and phosphorus--substances which are essential to life. Even in purely quantitative terms protein's presence is quite impressive. We are 18 to 20 percent protein by weight! Just as cellulose provides the structural framework of a tree, protein provides the framework for animals. Skin, hair, nails, cartilage, tendons,

---

From *Diet for a Small Planet* by Frances Moore Lappe. Reprinted by permission of the author's agent.

muscles, and even the organic framework of bones are made up largely of fibrous proteins. Obviously, then, protein is needed for growth in children. But it is also needed by adults to replace tissues that are continually breaking down and to build tissues, like hair and nails, that continue growing.

Furthermore, the body depends on protein for the myriad reactions that we group under the heading "metabolism." As regulators of metabolic processes, we call certain proteins "hormones," and as catalysts of important metabolic reactions we call other proteins "enzymes." In addition, hemoglobin, the critical oxygen-carrying molecule of the blood, is a protein.

Not only is protein necessary to the basic chemical reactions of life, it is also necessary to maintain the body environment, so that these reactions can take place. Protein in the blood helps to prevent the accumulation of either too much base or too much acid. In this way it helps maintain "body neutrality," essential to normal cellular metabolism. Similarly, protein in blood serum participates in regulating the body's water balance--the distribution of fluids on either side of the cell membrane. (The distended stomachs of starving children are the result of protein deficiency, a state which allows fluid to accumulate in the interstitial spaces between the cells.)

Lastly, and of great importance, new protein synthesis is needed for antibody formation to fight bacterial and viral infections.

Not only do we need protein for all of these vital body processes but we need to renew our body's supply every day. Whereas it takes from a few days to seven years to deplete the body's reserves of other required nutrients, amino acid reserves are depleted in a few hours.

So we need protein, but two basic questions still face us: how much and what kind? Since the answer to the question, "how much?," depends, in part, on "what kind?" I'll first explore the criteria by which we can distinguish among dietary proteins.

B. QUALITY MAKES THE PRODUCT

If all proteins were the same there would be no controversy about preferable protein sources for humans--only quantity would matter. But proteins aren't identical. The proteins our bodies use are made up of twenty-two amino acids, in varying combinations. Eight of these amino acids can't be synthesized by our bodies; they must be obtained from outside sources. These eight essential amino acids (which I will refer to as EAAs) are <u>tryptophan</u>, <u>leucine</u>, <u>isoleucine</u>, <u>lysine</u>, <u>valine</u>, <u>threonine</u>, the <u>sulfur-containing amino acids</u>, and the <u>aromatic amino acids</u>.

To make matters more difficult, our bodies need each of the EAAs <u>simultaneously</u> in order to carry out protein synthesis. If one EAA is missing, even temporarily, protein synthesis will fall to a very low level or stop altogether.

And to complicate things further, we need the EAAs in differing amounts. Basically, the body can use only <u>one</u> pattern of the EAAs. That is, each EAA must be present in a given proportion. In most food proteins all of the EAAs are present, but unfortunately one or more of the EAAs is usually present in a disproportionately small amount, thus deviating from the one utilizable pattern. These EAAs are quite rightfully called the "limiting amino acids" in a food protein.

Let us put together these three critical factors about protein:
Of the twenty-two necessary amino acids, there are eight that our bodies cannot
    make but must get from outside sources.
All of these eight must be present simultaneously.
All of these eight must be present in the right proportions.

What does this mean to the body? A great deal. If you eat protein containing enough tryptophan to satisfy 100 percent of the utilizable pattern's requirement, 100 percent of the leucine level, and so forth, but only 50 percent of the necessary lysine, then, as far as your body is concerned, you might as well have eaten only 50 percent of <u>all</u> the EAAs. Only 50 percent of the protein you ate is used <u>as</u> protein and the <u>rest</u> is literally wasted. The protein "assembling center" in the cell uses the EAAs at the level of the "limiting amino acid" and releases the left-over amino acids to be used by the body as fuel as if they were lowly carbohydrates. Chart III gives you a graphic illustration of what this means.

One reflection of how closely the amino acid pattern of a given food matches that which the body can use is what nutritionists term the "biological value" of a food protein. Roughly, the "biological value" is the proportion of the protein absorbed by the digestive tract which is retained by the body. In other words, the biological value is the percentage of absorbed protein that your body actually uses. But there's another question: How much gets absorbed <u>to begin with</u> by the digestive tract? That's what we call digestibility. So the protein available to our bodies depends on its biological value <u>and</u> its digestibility. The term covering both of these factors is <u>Net Protein Utilization</u> or NPU. Quite simply, NPU tells us how much of the protein we eat is actually available to our body (see Chart IV).

NPU is a key concept used throughout the remainder of the book, so it is important to become completely comfortable with the term. Let's take another look at what determines the NPU of a given food protein. The NPU of a food is largely determined by how closely the essential amino acids in its protein match the body's one utilizable pattern. Because the protein of egg most nearly matches this ideal pattern, egg protein is used as a model for measuring the amino acid patterns in other food. Let's take an example. A glance at Chart V will tell you that the amino acid pattern of cheese nearly matches egg's pattern while that of peanut fails utterly. You can guess then that the NPU of cheese is significantly higher than that of peanuts. The difference is great--70 as compared to about 40.

Prepared with an understanding of the important differences among food proteins, let's now turn to the second basic question:

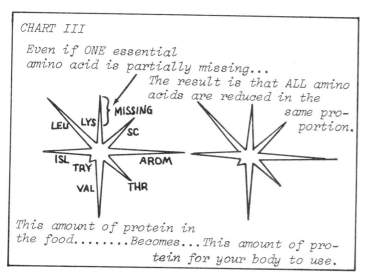

CHART III

Even if ONE essential amino acid is partially missing...

The result is that ALL amino acids are reduced in the same proportion.

This amount of protein in the food........Becomes...This amount of protein for your body to use.

C. <u>HOW MUCH IS ENOUGH?</u>

Dissatisfaction with the national diet and its fixation on protein has led some Americans to completely reject the need for even a minimum daily protein intake. The <u>Berkeley Tribe</u> recently published an article reflecting the danger of such an overreaction:

Several cases of <u>kwashiorkor</u> (severe protein malnutrition), a disease native to North Africa, (sic) have been found in Berkeley.

An unpublished UC [University of California] hospital report blames certain fasting, vegetarian, and especially macrobiotic diets for this. Those diets often re-

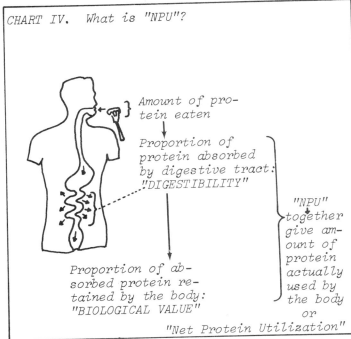

CHART IV. What is "NPU"?

- Amount of protein eaten
- Proportion of protein absorbed by digestive tract: "DIGESTIBILITY"
- Proportion of absorbed protein retained by the body: "BIOLOGICAL VALUE"
- "NPU" together give amount of protein actually used by the body or "Net Protein Utilization"

sult in clinically protein-deficient people

Other ailments caused by protein-deficient diets are wound infections and poor healing ability.

This report underscores the importance of knowing the facts of protein nutrition before experimenting with a new diet.

So far we have talked about the two extremes--overconsumption and protein deficiency; but just how much protein is _enough_? We can arrive at a satisfactory answer. (Although, as you might expect, the experts disagree.) Determination of the proper protein allowance for a population involves three separate considerations: (1) minimum need; (2) an allowance for individual differences; and (3) an adjustment for protein quality. Fortunately, disagreement among nutritionists is limited to only the first consideration, the body's minimum need for protein. And even here the range of differences is small enough to make an average meaningful.

1. _Minimum Need_ Since nitrogen is a characteristic and relatively constant component of protein, scientists can measure protein by measuring nitrogen. To determine how much protein the body needs, experiementers first put subjects on a protein-free diet. They then measure how much nitrogen is lost in urine and feces. They add to this an amount to cover the small losses through the skin, sweat, and internal body structure. And, for children, additional nitrogen for growth is added. The total of these nitrogen losses is the amount you have to replace by eating protein. It is therefore the basis of the minimum protein requirement for body maintenance.

Since the major expert bodies on nutrition arrive at different conclusions using this "factorial" method, no established minimum now exists to guide us. Consequently, we must be satisfied with a base line which averages the values proposed by three major expert groups, the Food and Nutrition Board of the (U.S.), National Academy of Sciences, the Canadian Board of Nutrition, and the Food and Agriculture Organization of the U.N. (I selected these particular groups because they are representative of the range of opinion among the many national bodies with protein standards.) The resulting recommended level of minimum intake is 0.214 grams per pound of body weight per day, which is close to the 0.227 gram level recommended by the National Academy of Science.

2. _Allowance for Individual Differences_ An allowance must also be made for individual differences. Fortunately most experts agree that a 30 percent allow-

ance will cover 98 percent of a population.* Adding an allowance of 30 percent to our minimum requirement of 0.214 grams gives us an allowance of 0.278 grams, or <u>0.28 grams per pound of body weight per day</u>.

3. <u>Adjustment for Protein Quality</u>. Finally we must take into consideration the <u>kind</u> of protein eaten. Here is where we can apply our understanding of protein quality. Recall that the basic distinction among food proteins is how completely your body can use them. Because your body can't use <u>low</u>-quality protein as completely, it stands to reason that you must eat <u>more</u> of a low-quality protein than a high-quality protein to fill a daily protein allowance.

a. <u>Allowance Based on Total Protein</u>. But the protein allowance we've discussed so far holds only for the highest-quality protein, one that would be used <u>completely</u> by the body. Since the total grams of protein eaten are <u>never</u> fully usable by the body, the task for nutritionists is to set appropriate protein allowances for different population groups, depending on the average usability of the protein (NPU) characteristic of that national diet. That is, the grams of <u>total</u> protein recommended must be increased to take into account the fact that not all the protein consumed can be used by the body. The formula for arriving at this allowance for grams of total protein is quite simple: See Table 1.

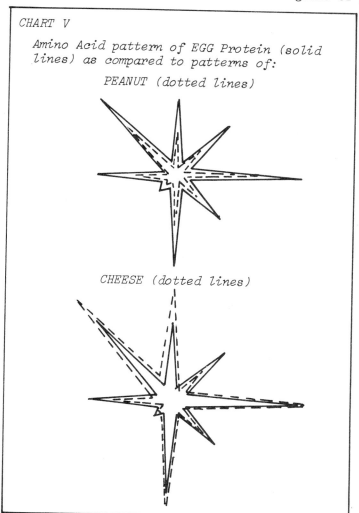

CHART V

Amino Acid pattern of EGG Protein (solid lines) as compared to patterns of:

PEANUT (dotted lines)

CHEESE (dotted lines)

Using this formula it is easy to determine allowances for total protein suitable for different types of national diets--diets of high-quality protein needing fewer grams of total protein and diets of lower-quality protein needing more grams. For example, since 70 is the average value for protein quality (NPU) of a diet based largely on animal protein (meat, egg, milk), we insert 70 into our formula (0.28 grams

* The Food and Agriculture Organization differs slightly but with comparable results. They add 10 percent to the minimum requirement for losses due to stress and an allowance of 20 percent for individual differences.

X 100/70 = 0.40 gram). The result is an allowance of 0.40 gram of total protein per pound of body weight per day. For example, since the average American woman weighs 128 pounds (and consumes largely animal protein), her daily allowance is 51 grams of total protein, or 0.40 X 128 pounds. (If you have difficulty multiplying by 0.40, just double the weight, divide by five and you have the answer expressed in grams.)

Table 1.

Protein allowance if eating fully usable protein (0.28 grams per pound of body weight)	X	100 / Net Protein Utiliztion characteristic of the national diet	=	Grams of total protein recommended for that population (per person/per pound body weight/per day)

Now we are on solid ground. And we are in a position to evaluate the allowance set for us as a population by the National Academy of Sciences. It is set at 0.42 gram per pound of body weight per day. At 0.42 gram it is 5 percent above the 0.40 gram which we have seen to be appropriate for a population getting most of its protein from animal sources. So now when a doctor tells you that you need 65 grams of protein a day, as an average male, or 54 grams, as an average female, you'll know where the figure comes from. The weight of the average American male and female (154 pounds and 128 pounds, respectively) is simply multiplied by 0.42 gram, the factor recommended by the National Academy of Sciences.

Consider now a diet based largely on plant protein. Since a typical value for the quality of plant protein is 55, a higher total protein allowance is called for. Inserting 55 into our formula, the result is 0.51 gram of total protein per pound of body weight per day (0.28 gram X 100/55 = 0.51 gram). Thus the recommended allowance for a 128-pound person in a population on a largely plant protein diet is 65 grams of total protein, or 0.51 X 128 pounds. (Notice that 0.51 is close to one-half; thus you can quickly estimate the grams of total protein needed for a person on a largely plant protein diet by dividing the body weight in half and expressing the answer in grams.)

b. <u>Allowance Based on Usable Protein</u>. But in this book where each food is listed individually, we can go one step further than generalizations about the protein quality in the diet of a whole population or an individual. In fact there is no need to generalize and to obscure the wide differences in protein quality of a mixed animal and plant protein diet. In such a diet only one-third of the protein from some sources is usable by your body while practically all the protein from other sources is usable. In the protein tables that make up Part III I have accounted for these differences by simply <u>adjusting the protein in each food to the level that is fully usable by the body</u>. Since each food is adjusted for protein quality, there is no need to apply the earlier formula based on an <u>average</u> level of protein quality. So, instead of talking about grams of total protein (only part of which the body can use), I'll be talking about grams of <u>usable</u> protein. Recall that the recommended daily protein allowance based on usable protein is 0.28 gram per pound of body weight, or 35.8 grams of usable protein a day for a person weighing 128 pounds and 43.1 grams for a person weighing 154 pounds.

This makes much more sense and will become even clearer to you in Part III where I explain how to use the protein food tables. But before reading on I'm sure you would like to have some idea of what the recommended intake of usable protein means in terms of the food you eat. Here are some comparisons, based on the hypothetical assumption that you would be getting all of your day's protein from one source:

If you weigh:	To meet a day's allowance of usable protein:	YOU NEED					
		meat or	fish or	milk or	eggs or	dry beans or	nuts
128 lb	35.8 g	7 1/3 oz	8 1/3 oz	5 cups	6	12 3/4 oz	12 oz
154 lb	43.1 g	9 oz	10 oz	6 cups	7	15 1/3 oz	14 1/3 oz

D. PROTEIN INDIVIDUALITY

Before going on, a word of caution about putting too much stock in any figures purporting to deal with the "average" human being. R. J. Williams, a nutritionist who has devoted himself to the study of individual nutritional differences, illustrates dramatically the range of our "protein individuality." He points out that if beef were the only source of protein, one person's minimum protein needs could be met by two ounces of meat; yet another individual might require eight ounces. Although over 98 percent of a population may not range more than 30 percent from an average requirement, these two possible extremes represent a four-fold difference in protein requirement! And, requirements for other nutrients are found to be equally, or even more widely, disparate.

Even more surprising perhaps is the fact that the need for protein can vary within the individual. Certain physical stress (pain, for example) or psychological stress (even from exam pressure) can cause one's protein requirement to jump by as much as one-third over ordinary needs.

The obvious conclusion is this: The fact that your friend is thriving beautifully on a low-protein diet tells you nothing about a diet suited to your own body's needs. The best answer is to develop what Dr. Williams calls "body wisdom" which involves more than just being aware of how you feel--your energy level, general health, and temperament. Certain nutritional deficiencies have been shown to negatively affect one's appetite and choice of foods; so just feeling "satisfied" is not enough. Part of "body wisdom" involves being a wise observer of your body's condition. Because nails, hair, and skin require newly synthesized protein for growth and health, their condition is usually a good indication of whether or not you're getting enough protein. Similarily, notice whether or not abrasions heal quickly. If they don't, you may be seriously lacking protein in your diet.

Now that we can estimate the amount of protein that human beings must have (and understand how, in part, it depends on the type of protein eaten), we're ready to consider the really practical question: What are the best protein sources and how can we make best use of them? Since there is a great deal of "mythology" surrounding protein sources, let's first get our thinking straight about the useful distinctions to be made among them.

E. IS MEAT NECESSARY?

Those who insist on the superiority, in fact indispensability, of meat as a protein source base their argument on both the large quantity and the high quality of protein in meat. Plant protein is seen as inferior on both counts. The result is that animal and vegetable protein are thought of as comprising two separate categories. In fact, this is a common mistake in our thinking about protein. For our nutritional concern here it is much more useful and accurate to visualize animal and vegetable protein as being on one continuum.

Chart VI, "The Food Protein Continuum," will help you see the range of protein variability on two scales: protein quantity, based on the percent of protein in the food by weight; and quality, based on the NPU or usability of the protein by the body.

Quantity: When judging protein with quantity as the criterion, generalization is difficult. However, it is clear that plants rank highest, particularly in their processed forms. Soybean flour is over 40 percent protein. Next comes certain cheeses such as Parmesan which is 36 percent protein. Meat follows, ranging between 20 and 30 percent. Dried beans, peas and lentils are essentially in the same category, that is, between 20 and 25 percent protein. At the bottom end of the quantity scale we find examples of both animal and plant protein. We find grains here and, though it might surprise you, milk and eggs also. There are, of course, other plants, some fruits, for example, that contain too little protein to even appear on the scale. (We're concerned here only with plants that are widely used as sources of protein.)

Quality: The protein quality scale generally ranges from NPU values of about 40* to 94. Clearly animal protein occupies the highest rungs of this scale. Meat, however, is not at the top. It places slightly above the middle with an average NPU of 67. At the top are egg (NPU of 94) and milk (NPU of 82). The NPUs

*Most of the NPU values used throughout the book are taken from a United Nations publication.

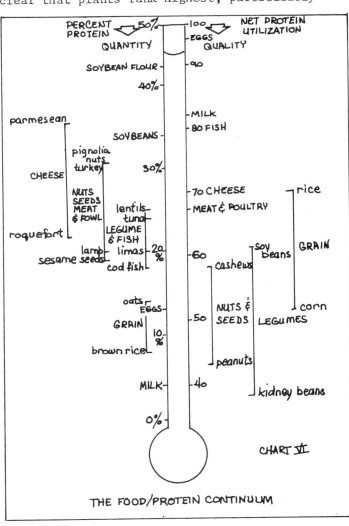

THE FOOD/PROTEIN CONTINUUM

of plant proteins generally range lower on the continuum, between 40 and 70. However, protein in some plants such as soybeans and whole rice approach or overlap the NPU values for meat. But the general distribution of animal protein high on the NPU scale and plant protein lower on this scale tells us that the proportions of essential amino acids found in most animal protein more nearly match human body requirements than the proportions commonly found in plants. This means that in general you need to eat proportionately less meat protein than plant protein to be "covered" for the essential amino acid requirements.

But people don't have to depend on meat for protein and the correct supply of amino acids. There are several other alternatives:

1) Eat large amounts of "lower-quality" plant protein, enough so that you will get an adequate amount of even the "limiting amino acids."
2) Eat alternate animal protein sources such as dairy products.
3) Eat a variety of plant proteins which have mutually complementary amino acid patterns.

When eating plant protein from a single plant source (such as beans or rice) you're likely to be limited in the amount of protein your body can utilize because of a limiting amino acid (for beans the limiting amino acids are the sulfur-containing amino acids and for rice they are isoleucine and lysine). Consequently, a major drawback of the first alternative is that you would have to eat (and waste!) relatively enormous quantities of your protein source in order to insure your daily protein requirement.

The advantage of the second choice is that dairy products have high-quality protein, in fact better values for protein utilization (NPU) than meat. But by itself, of course, this alternative is gastronomically dull; and moreover, since protein conversion even by dairy cattle entails some waste of protein, why rely on this alternative more than you have to?

The third alternative means eating, in the same meal, different plant foods in which the amino acid deficiency of one item is supplemented by the amino acid contained in others. (Remember that the EAAs must be present simultaneously.) It is more efficient than the first alternative because the complementary effect of the mixture means that more of the protein can be used by the body (less is lost and converted to fuel). And it is more efficient than the second choice because it takes optimal advantage of more abundant plant protein.

F. COMPLEMENTING YOUR PROTEINS

Obviously the best solution is to use both the second and third alternatives. This means combining different plant sources, or nonmeat animal protein sources with plant sources, in the same meal. Most people do this to some extent anyway, just as a matter of course. Eating a mixture of protein sources can increase the protein value of the meal; here's a case where the whole is greater than the sum of its parts. To repeat, this is true because the EAA deficiency in one food can be met by the EAA contained in another food. For example, the expected biological value of three parts white bread and one part cheddar cheese would be 64 if there were no supplementary relationship. Yet, if eaten together, the actual biological value is 76! The "whole" is greater largely because cheese fills bread's lysine and isoleucine deficiencies. Such protein mixes do not result in a perfect protein that is fully utilizable by the body (remember that only egg is near perfect). But combinations can increase the protein quality as much as 50 percent above the average of the items eaten separately.

Eating wheat and beans together, for example, can increase by about 33 percent the protein actually usable by your body. Chart VII will help you see why. It shows the four essential amino acids most likely to be deficient in plant protein. On each side, where beans and wheat are shown separately, we see large gaps in amino acid content as compared to egg protein. But, if we put the two together, these gaps are closed.

To exploit this complementary effect, you can make dishes and plan meals so that the protein in one food fills the amino acid deficiencies in another food. A bit laborious, you say? It's not as hard as it sounds! And to prove it, I've included many recipes* to guide (and tempt) you. But the real fun for you might be "inventing" your own complementary protein combinations.

G. PROTEIN ISN'T EVERYTHING

Many people who might otherwise rely more on plant sources for protein continue to eat great quantities of meat because they believe that only "good red meat" can supply the many vitamins and minerals that their bodies needs. Are they right?

A national food survey in Britain in 1966 showed that, although 40 percent of the protein in a typical British diet comes from plant sources, <u>plants provide, on the average, more than twice the amount of vitamins and minerals provided by meat and fish</u>. Seven nutrients were considered: vitamin A, thiamine, riboflavin, niacin, vitamin C, calcium, and iron. Plant sources were the greatest contributors of all nutrients except riboflavin and calcium. But dairy products and eggs, <u>not</u>

* See Part IV, Section B.

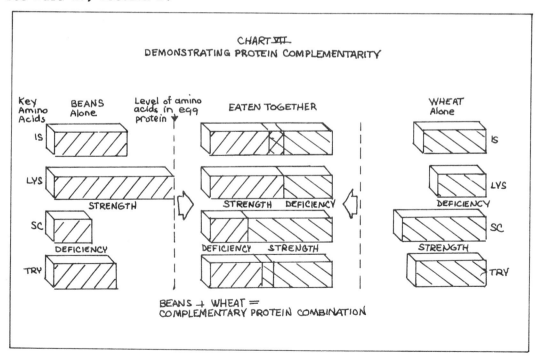

meat, provided the bulk of the daily requirements of these two nutrients. This general pattern emerged in spite of the fact that green vegetables, which are valuable sources of many of these nutrients, comprise a relatively minor part of the British diet.

A breakdown of sources of vitamins and minerals in the American diet reveals a similar pattern. Although plant sources provide only 31 percent of the protein in our national diet, they provide 50 percent of the vitamin A, 59 percent of the thiamine, 53 percent of the niacin, 94 percent of the vitamin C, and 62 percent of the iron. As in the British study, dairy products contribute the largest percentage of both riboflavin and calcium.

Other important nutrients not covered by the British survey include phosphorus, potassium, and magnesium. Although meat is a good source for both phosphorus and potassium, there are nonmeat sources which are even better. Whereas meat and fish contain 250 to 274 milligrams (mg) of phosphorus per 100 grams (3 1/2 ounces), cheddar cheese contains 478 mg and peanuts 401 mg. Meat contains from 290 to 390 mg of potassium per 100 grams, but a baked potato has 503 mg and lima beans 422 mg. In the case of magnesium, meat is actually among the poorer sources. Rich sources include cocoa, nuts, soybeans, whole grain, and green leafy vegetables.

In fact, the only required nutrient thought to be limited strictly to animal sources is cobalamin (vitamin B_{12}). But there are potentially significant exceptions to this rule. A type of blue-green algae, Spirulina maxima (a popular food in parts of Africa), and two other algae all contain cobalamin. Thus, it is possible, at least theoretically, to fill all man's nutritional needs from plant sources. For most of you, however, the important fact is that there is no danger of cobalamin deficiency as long as you eat dairy products or eggs.

Thus, we can safely conclude that a varied plant protein diet supplemented with dairy products and eggs can supply sufficient protein while at the same time surpassing meat in the provision of some of the other basic nutrients. All this is not meant to belittle the nutritional value of meat. My aim is only to provide a more realistic view of the wide variety of nutritious foods sources to replace the culturally fixed idea of the absolute supremacy of meat.

*

PEANUT BUTTER PROTEIN

As soon as they pass through the automatic sliding glass doors, parnoia appears. Necks stiffen, hands clutch purses, eyes search desperately for bargains. A trip to the local supermarket is no longer an adventure in good eating; it is a survival test. And in the current food crisis it is the careful eater who is faring worst. Food that is empty of nutrients or loaded with chemicals is cheap; nutritious food is expensive.

During the past few months, reports of record increases in food prices have been accompanied by government advice on how to eat well without consuming your entire paycheck. At a White House news conference, Virginia Knauer, special assistant to the president for consumer affairs, delivered a nine-point "Battle Plan for Saving," including a suggestion that Americans begin eating the hearts, liver, and kidneys instead of the breasts, loins, and ribs of their favorite farm animals. Perhaps the most revolutionary idea came from James D. Maclane, deputy directory of the Cost of Living Council, who suggested that Americans "try something that I've tried lately--eat a little less."

Not everyone in the administration has lost his appetite, however. The U.S. Department of Agriculture recently published a helpful guide to cutting meat expenditures without nutritional loss. The report, which appeared in Family Economics Review, compares the cost of protein from various meats and meat alternates, at August 1972 prices. Twenty grams of protein, about one-third the recommended daily allowance for most adults, can be obtained from a three-ounce serving of such lean meats as beef, pork, lamb, veal, turkey, or fish. To get that 20 grams from sirloin beefsteak, you will pay 59 cents. The same 20 grams from turkey,

Copyright © 1973 by Saturday Review Co. Used with permission.

however, costs only 22 cents.

Moreover, as the guide shows, not all meats and meat alternatives provide 20 grams of protein in a three-ounce serving. Twenty grams of protein requires six ounces of bologna, ten slices of bacon, one-half pound of pork sausage, or three and a half frankfurters. In dollars and cents this means you pay 46 cents for 20 grams of protein from bologna, but only 19 cents for the same amount of protein from hamburger, even though, on the average, hamburger costs 13 cents more per pound. Actually, the most economical protein sources are meat alternatives like dry beans, eggs or peanut butter. Four and a half tablespoons of peanut butter, for example, provides 20 grams of protein and costs only 12 cents.

The Department of Agriculture apparently does not see a quick end to the food crisis. The Cost of Living Council agrees. The "rosier outlook" is that price increases will be slower in the latter half of 1973. So, if you're not looking forward to a steady diet of peanut butter sandwiches, you can send for a copy of the USDA's report by writing to the Consumer & Food Economics Institute, Federal Center Building #1, Hyattsville, Maryland 20782.

STRICTER CONTROLS ON VITAMINS

Our lack of conclusive knowledge about the benefits and/
or ill effects of vitamins is a paradigm of the undeveloped
state of the science of nutrition. The following three articles
deal with some effects of vitamins, and the legal battle which
has developed over the sale of vitamin concentrates.

A battle over controls on the sale of food supplements now is joined--and it promises to be a bitter one.
 The Food and Drug Administration on August 1 ordered strick limits on the labeling, promotion and sale of vitamins and minerals. Its action came after 10 years of study and debate over the highly controversial issue.
 The decision, to become effective October 1, immediately raised cries of protest from health-food enthusiasts, some drug manufacturers and those persons who claim that massive doses of vitamins are the answer to various problems connected with health.

WHAT'S REQUIRED

Included in the package of 19 regulations, proposals and policy directives is one that will require prescriptions for high-dosage forms of vitamins A and D. Smaller units of the same vitamins will continue to be readily available over the counter.

Reprinted from *U.S. News & World Report*. Copyright 1973, U.S. News & World Report, Inc.

Other regulations set strict labeling rules. They aim to prevent what the FDA called "consumer deception" through sales of such products as fruit drinks that may contain no real fruit juice or "main dish" products lacking the principal ingredient, such as meat.

Dr. Alexander Schmidt, FDA Commissioner, said of the decision:

"The opposition with which we are most concerned stems from the honest fears of many citizens. Some fear that FDA is going to make certain vitamin pills unavailable or, if available, then only by prescription and at higher cost. Others fear that FDA may infringe on their right to decide what they will eat.

"None of this is true. The single and most important purpose and effect of the regulations is to require full and honest labeling and fair promotion of vitamin and mineral products as the basis for a more-informed consumer choice.

"The new regulations are based on the best and broadest scientific evidence and expert advice."

WHAT'S PROMISED

Dr. Schmidt pledged that the regulations will not ban any vitamin or mineral product from the market or drive out of business any manufacturer willing to provide proper formulation and labeling of his product.

The FDA said it will allow any vitamin and mineral product containing 50 per cent of the new standard--called the Recommended Daily Allowance--to be marketed as a food. Any product containing between 50 and 150 per cent of the allowance will be regarded as a dietary supplement, and any exceeding 150 per cent will be designated as an over-the-counter drug subject to later review.

A normal, well-balanced diet should provide 100 per cent of the Recommended Daily Allowance for the average adult, the officials said.

COURT BATTLE AHEAD?

The National Health Federation, which represents thousands of health-food enthusiasts, vigorously challenged the regulations.

"There is no question about the fact that we will go to court," said Clinton Miller, the group's vice president.

Mr. Miller said the Federation also will pursue efforts to get Congress to overturn the rules. It claims 165 sponsors in the House of Representatives for a bill to prohibit FDA restrictions on vitamins and minerals unless a safety threat can be demonstrated.

Legislation to free vitamins from FDA controls was introduced in the House months ago, but it has been stuck in a subcommittee ever since.

WHAT VITAMINS, ARE

Vitamins are chemical compounds essential to health that are present in different quantities in all unprocessed foods. Some 20 vitamins are also available in pure form such as pills, powders or solutions, or as additives to foods and beverages.

Nutrition scientists are not sure just how all the vitamins work or what they do in the body. They are reasonably certain, however, of the operation of four categories of vitamins:

<u>Vitamin A</u>--carotene. Essential in formation of body tissue, eye coloring and teeth. Lack of vitamin A can cause night blindness, damage to digestive tract, bones, reproductive tract, skin and teeth in children. Dangers from overdose: kidney damage can lead to death. Natural sources: fish and animal flesh, particularly liver.

<u>Vitamin B Group</u>. At least 12 different vitamins are lumped together in the vitamin B complex. The most important ones are thiamine, riboflavin, niacin, folic acid and vitamin B-12. They help in blood formation. Deficiency dangers: anemia, pernicious anemia, heart and stomach trouble, headache and fatigue. There are few apparent dangers of overdose. Natural sources: meat, fish, yeast, eggs and cereals.

<u>Vitamin C</u>--ascorbic acid. Forms the "cement" and "mucillage" in bones, teeth and soft tissues of the body. Some doctors believe that massive doses will ward off, or alleviate, symptoms of the common cold. Deficiency dangers: scurvy, swollen and bleeding gums, loose teeth. Overdose dangers: under hot debate. Natural sources: citrus fruits, tomatoes, strawberries, cabbage, spinach.

<u>Vitamin D</u>--turns calcium and phosphorus in foods into bone. Lack of it in children causes rickets; in adults, porous, brittle bones. Natural sources: eggs, butter, fish-liver oils. Overdose dangers: can cause serious blockages in soft tissues of the liver and kidneys, sometimes leading to death.

*

MILK AND BLINDNESS IN BRAZIL
G. Edwin Bunce

MALNUTRITION is one of the most serious problems facing the world today. The Food and Agriculture Organization (FAO) in its Third World Food Survey estimated that at least 20 per cent of the population of the developing countries are undernourished and that about 60 per cent consume diets that are inadequate in nutritional quality. Despite considerable efforts to feed the hungry, we are failing. Today more persons are hungry than ever before in the history of the world.

Many laymen are aware of the basic facts of good nutrition and the need for adequate daily supplies of the various macro and micro nutrients. Most are unaware, however, of the relationships among nutrients or between nutrients and environmental factors. Moreover, conventional medical training and practice in this country often ignore many of these interdependencies, which affect markedly the physiology of the consuming organism. Such interactions are the rule rather than the exception.

In order to illustrate this point more fully, let us consider the case of the interrelationship of protein and vitamin A, two nutrients in deficient supply in many populations that have severe deficiency symptoms. Vitamin A is one of the class of fat soluble vitamins. It participates in a number of biochemical processes and is necessary for a normal rate of growth, for night vision, and for the health of epithelial structures of the eye, the respiratory tract, and the gastrointestinal tract. A mild deficiency of vitamin A may result in the inability to see in dim light, growth retardation, and a dermatosis characterized by dryness, roughening, and itching of the skin. If the deficiency is more severe, the conjunctiva and para-ocular glands become abnormal and unable to produce their usual supply of

Reprinted with permission from *Natural History* Magazine, February 1969. Copyright © The American Museum of Natural History, 1969.

lubricating fluids. This condition, called xerophthalmia, may be readily followed by secondary infection, ulceration, softening of the cornea (keratomalacia), and perforation. If therapy is begun at the stage of uncomplicated xerophthalmia, the prognosis is good. Otherwise, partial or total blindness is likely to result. Vitamin A deficiency is common in underdeveloped regions, and particularly prevalent in urban slums.

Protein is one of the macronutrients. Proteins are large polymeric molecules formed from smaller compounds called amino acids. Protein serves a structural role as the primary material of muscle and other cells and is required for the production of enzymes. Protein deficiency (kwashiorkor) and protein-calorie malnutrition (marasmus) are the two most widespread nutritional deficiencies in the world today. Failure to consume adequate amounts of protein and calories leads first to growth failure, followed by wasting, secondary infections, and death.

The relationship between deficiencies of vitamin A and of protein is complex. Animals fed a good protein diet apparently consume their body stores of vitamin A rapidly and develop eye lesions at an earlier time than slower-growing littermates fed rations low in protein. Clinical evidence with human infants similarly shows that children with grossly retarded growth often have no ocular lesions despite low vitamin A intakes and blood levels, but they may develop xerophthalmia if vitamin A supplements are not included with the treatment for protein malnutrition. The picture is further complicated by the fact that serum proteins act as carriers for vitamin A in the blood; thus, mobilization as well as utilization of stored vitamin A may be stimulated by protein supplements in children previously receiving low-protein diets.

With this information as background, one may speculate that prolonged intake of a marginal level of both vitamin A and protein could result in a juvenile population in which growth retardation was present, but not severe, and the body reserves of vitamin A were minimal. Sudden famine, brought on by drought, floods, revolutions, etc., would precipitate protein-calorie malnutrition, but the concurrent acute growth restriction or weight loss would prevent the outbreak of visible signs of vitamin A deficiency. Distribution of a high-quality protein low in vitamin A to children in this state would yield a spurt of growth and rapid mobilization and consumption of the remaining body stores of vitamin A. The acute stage of vitamin A deficiency with its potential for partial or total blindness might then follow.

The likelihood that such a series of events might occur was very real up to the early 1960's, since one of the major foods employed in famine relief was skim milk powder, a good-quality protein low in vitamin A. Relief agencies often insisted upon distributing vitamin A capsules along with milk powder, with instructions that the children might lose their sight if they did not take the capsules. However, the number of children who received the vitamin A, as suggested, is not known. One may presume it likely that these warnings were ignored or forgotten by many uneducated and destitute parents. However, few instances of widespread and seious complications are known. Perhaps most recipient populations were not in the state of balance required to be susceptible in large numbers. Perhaps the vitamin A capsules were consumed more faithfully than one might anticipate. On the other hand, the difficulties of obtaining reliable medical statistics may have obscured the true incidence.

In at least one instance, protein supplements seem to have aggravated chronic vitamin A deficit. In the late 1950's several severe droughts occurred in northeast Brazil, where deficits of both protein and vitamin A are serious, well established, and of long standing. Thousands of country people migrated to Recife and other cities seeking work. Epidemic outbreaks of xerophthalmia and keratomalaci were reported from Recife, coincident with distribution of dried skim milk to famine victims. Carefully documented evidence of this event is not available, but first reports have been supported by subsequent interviews with competent authorities.

Recognizing the potential hazard of separate consumption of vitamin A supplements and skim milk powder, UNICEF and other agencies encouraged research on additives that could be mixed with the milk powder. A stable, water-dispersible vitamin A derivative was developed by chemists in the early 1960's. Tests with rats showed the additive to be effective even with low fat intakes. However, the possibility still remains that humans in the affected area might respond differently because of unsuspected variations in conditions (such as parasite loads or chronic deficiencies), and it seemed advisable to study recipient subjects to assure safety and effectiveness under field conditions. Data from these studies supported the animal tests and encouraged the widespread employment of the additives in human feeding programs.

I believe that the agencies involved in the food relief programs for northeast Brazil made a reasonable effort to minimize the dangers of aggravated vitamin A deficiencies. The value of this example lies in its warning of the necessity for careful planning with nutrition experts before beginning projects dealing with food distribution. For instance, if an administrative decision had been made to economize by omitting vitamin A supplements altogether--on the grounds that few cases of xerophthalmia were being reported--it might have seemed quite sound to a person unfamiliar with the nutritional history of northeast Brazil and the interrelationship of these two nutrients. Yet, it could have caused immense suffering.

Nutrient and environment interdependencies are not rare in the study of nutrition. The failure to produce niacin deficiency in rats fed diets similar to those consumed by human pellagra patients was not understood until research revealed the rat to be about twice as efficient as the human in converting the amino acid tryptophan to niacin. Recent studies in Africa and the West Indies have shown that magnesium deficiency is frequently induced secondary to protein deficiency and must be considered in the therapeutic program for maximum success in treatment. Studies from South America and Tulane University indicate that the type of dietary carbohydrate may be of major significance in determining the response of a human to various parasitic species.

The interaction between vitamin A and protein resulted in the northeast Brazil experience of aggravating a vitamin A deficit by the feeding of high-protein supplement. Such interactions are not readily obvious. Since nutrient interrelationships are common and may have profound effects if ignored, the many factors in dietary patterns should be considered before changes are put into effect. At the least, this will be an attempt to minimize the potential for unsuspected and tragic results.

*

VITAMINS AND THE FETUS: THE BENEFITS OF B 12

Nutrition scientists interested in the well-being of unborn and newly born infants have focused largely on the effects of severe protein deficiencies. They have paid less attention to the effects of vitamin deficiencies. For these reasons Paul M. Newberne and Vernon R. Young of the Department of Nutrition and Food Science at the Massachusetts Institute of Technology decided to study maternal intake of vitamin B_{12} during pregnancy and its long-range effects on the newborn.

Vitamin B_{12} is one of the vitamins that animals and humans do not manufacture in their bodies but must take in with their foods. It is essential for a number of metabolic roles. It promotes growth in adolescence. It helps cells use DNA, the genetic material of life. It helps various enzymes, particularly liver enzymes, catalyze different reactions. Liver enzymes are one of the body's prime defenses against foreign toxic substances.

Newberne and Young mated rats, then gave some of the females a standard B_{12} diet, and others a somewhat higher B_{12} diet. Both groups received the same food during pregnancy. After giving birth, all the rats were put on the standard B_{12} diet. After their progeny were weaned, they two were given a standard B_{12} diet. This way the researchers were able to measure the effects of a higher maternal intake of B_{12} during pregnancy on the fetus, particularly as those effects carry over into the first months of life.

As Newberne and Young report in the March 23 Nature, they found that a higher maternal intake of B_{12} during pregnancy affected birth weight significantly. Pups born to mothers who had received more B_{12} weighed more than the other pups, and this greater weight continued during the first year of life. The pups whose mother

---

Reprinted by permission of *Science News*, © 1973 by Science Service.

had received more B_{12} also had more protein per body weight than did the other pups. The animals whose mother had had more B_{12} also showed more active liver enzymes, suggesting they might be better protected against infection. Indeed, the pups whose mother had received more B_{12} experienced lower mortality and more resistance to infection during the first months of life than did pups whose mothers had received less B_{12}.

The authors believe that these results might be extrapolated to the human situation. In other words, the B_{12} a woman consumes during pregnancy might affect the growth and health of her child, particularly if her vitamin B_{12} intake during pregnancy is marginal and her baby is subjected to trauma or disease before or soon after birth. "Questions about many of the unexplained illnesses in children and the wide variation among individuals in their resistance to disease," the authors declare, "may conceivably be answered by more intensive study of the prenatal nutrient needs of mother and her fetus."

But they caution pregnant women not to dose themselves with large amounts of any vitamin, particularly the B vitamins. Studies have suggested that the 15-odd B vitamins work synergistically, that is, as a complex, and if one B vitamin is consumed in much bigger quantities than the others, the effects can be harmful.

HOT DOGS AND HYPERKINESIS

The food processing industry has compounded the problems of the search for good foods.

Sitting down to a meal in 1973 often means an encounter not only with meat and vegetables but with scores of artificial flavors and colors as well. These chemical additives have no nutritional value and serve only to enhance the appearance and taste of processed foods like hot dogs, ice cream, soft drinks and ready-to-eat cereals--precisely the stuff consumed in enormous quantities every day by American children. In the past, these synthetic ingredients have been classified as safe by the Food and Drug Administration. But last week a California allergist told the AMA meeting that these additives may in some cases trigger the symptoms of extreme hyperactivity and severe learning disorders in schoolchildren.

Because artificial flavors and colors have been shown to cause a wide variety of allergic reactions, Dr. Ben F. Feingold of the Kaiser-Permanente Medical Center decided to test the possibility that the substances might also be responsible for some cases of hyperkinesis. Hyperkinesis is a behavioral disorder characterized by excessive physical activity and the inability to concentrate and learn. Working with some 25 hyperactive San Francisco schoolchildren, Feingold and his colleagues discovered that a number of the youngsters they tested had a history of other allergies, most had normal or high IQ's--and all of them ate large quantities of processed foods.

Reprinted from *Newsweek* Magazine, July 9, 1973. Copyright Newsweek, Inc. 1973; reprinted by permission.

Control: The children were placed on a diet eliminating all foods containing artificial flavors and colors. Within a few weeks fifteen of them showed dramatic improvement. One 7-year-old boy, for example, had been extremely hyperactive for several years. "When he was at home," Feingold recalls, "he stomped around, slamming the doors and kicking the walls and even charging oncoming cars with his bicycle." At school, his hyperactivity prevented him from learning and disrupted the rest of the class. A round of pediatricians, neurologists and psychiatrists could do nothing for him--until he was placed on Feingold's diet.

After a few weeks of dietary control, reports the allergist, the boy settled down and was able to perform well in school and behave properly at home. Interestingly enough, notes Feingold, any infraction of the diet led almost immediately--within a matter of hours--to a return of the hyperkinetic behavior. "We can turn these kids on and off at will," he says, "just by regulating their diet."

Because he has not yet done a controlled study, Feingold cautions that his observations must be regarded as preliminary. But in the meantime, he thinks that many cases of hyperkinesis can be controlled by a diet free of artificial flavors and colors. Unfortunately, keeping a child on the restricted regimen can be an all-but-impossible task. "Artificial colors and flavors are contained in 90 per cent of processed foods," notes Feingold. "Is it any wonder that our children are jumping and failing to learn?"

FOOD LABELS TO SAY MORE ABOUT NUTRITION

Labels on cans and packages of food are required to list contents according to quantity. Whatever is in a product in the largest quantity must be stated first, whatever is present in the next largest amount must be stated second, and so on. This has been the consumer's best guide to the nutritional value of what he buys. The Food and Drug Administration has now announced new regulations that will require all products that boast of nutritional or diet value to list specific nutritional and diet information on their labels.

If a product is said to be "enriched" or "fortified," for example, it must state serving size, servings per container, caloric content, protein content, carbohydrate content, fat content, percentage of U.S. Recommended Daily Allowances of protein, vitamins and minerals. If a product is a purported diet food, its label must show saturated and unsaturated fat content and cholesterol content per serving. Labels must be changed by the end of the year.

Reprinted by permission of *Science News*, © 1973 by Science Service.

BIRTH CONTROL: CURRENT TECHNOLOGY, FUTURE PROSPECTS
Jean L. Marx

Ours is a pill-taking society, and the use of drugs which can prevent conception is regarded with general social acceptance. "The Pill" and other contraceptive drugs will undoubtedly have lasting (and, perhaps, unexpected) effects on our future.

The news about birth control is that there is no news--at least, no news of the imminent availability of methods that differ radically from existing techniques for controlling human fertility. New variations on old themes, however, may offer better efficacy, more convenience, greater freedom from hazardous or uncomfortable side effects, or all of these. Advances in basic research on reproductive physiology also suggest that new techniques may be developed in the future--but a minimum of 10 to 15 years could be required before they are available for routine use.

The "pill", introduced in the early 1960's, did revolutionize birth control technology. Because the oral contraceptives produce virtually 100 percent inhibition of female fertility, their superior efficacy has not been questioned. Nevertheless, reports of side effects that range from the merely uncomfortable--nausea, excess water retention--to the potentially dangerous--a higher incidence of abnormal blood clots in users--have sparked efforts to formulate oral contraceptives without these disadvantages. One such effort is the "mini-pill," now being marketed by Ortho Pharmaceutical Corporation, Raritan, New Jersey, and by Syntex Corporation, Palo Alto, California.

Reprinted by permission from *Science*, Vol. 179, pp. 1222-1224, March 23, 1973. Copyright 1973 by the American Association for the Advancement of Science.

Oral contraceptives depend on synthetic steroids for their effectiveness. (Synthetic steroids must be used because they are not destroyed by the body's enzymes before they reach their target organs.) The older "pills," which contain both an estrogen and a progestogen, act primarily by inhibiting the monthly release of the egg from the ovary. The "mini-pill," on the other hand, contains only a progestogen in a daily dosage roughly one-third or less that of the other "pills"; the low concentration of progestogen apparently prevents the sperm from reaching the oviducts, where fertilization occurs, by maintaining the mucus at the opening to the uterus in a condition that hinders sperm migration.

Although most of the side effects of the "pill" are associated with the estrogenic component, the Food and Drug Administration (FDA) has warned that not enough data are available at present to determine whether the risks of blood-clotting are indeed lower with the "mini-pill." The FDA points out that a small percentage of the progestogen in the "mini-pill" is actually converted to an estrogen in the body. Moreover, the risk of pregnancy--almost 3 percent--is higher for "mini-pill" users. (A failure rate of 3 percent means that if 100 women use a contraceptive technique for 1 year, 3 of them will become pregnant.)

Other research on steroidal control of female fertility is directed at the design of more convenient methods of drug administration, especially those applicable in areas or countries where conventional medical care is not readily available. Some of these delivery methods use a plastic material impregnated with the contraceptive steroid, usually a progestogen. The plastic can be implanted under the skin or it can be fashioned into a ring that is inserted into the vagina. Depending on the amount of hormone released per day, steroids thus administered act either as ovulation inhibitors or by the same mechanism as the "mini-pill." The vaginal rings are worn for approximately 1 month and are then removed so that menstruation can occur. Subcutaneous implants, which are easy to remove if pregnancy is desired, can provide protection for a year or longer. The Upjohn Company, Kalamazoo, Michigan, has also been testing a long-active injectable progestogen called Depo-Provera in Europe; a single intramuscular injection can provide contraceptive action for 3 months. The steroid has not been approved for use as a contraceptive in this country because it produced breast cancer in dogs.

The intrauterine device or IUD was another major contributor to the birth control revolution of the 1960's. The original devices suffered from several liabilities--including a high failure rate of approximately 20 percent--and side effects such as cramping, and excessive or irregular bleeding. They were not well tolerated by women who had never had a baby. Recent modifications have conquered or at least minimized these problems. Intrauterine devices with chemical adjuncts have proved particularly effective. One such device consists of a plastic "T" with a copper wire wound about it. The copper, leached from the wire by uterine secretions, probably acts by preventing implantation of the fertilized egg.

AN IUD WITH PROGESTERONE

The IUD may also be used to deliver natural--not synthetic--progesterone directly to the target organ, the uterus. Alza Corporation, Palo Alto, California, has been testing such a device in more than 2000 women for 2 years with only one recorded pregnancy. According to Bruce Pharriss of Alza, the polymeric film that coats the T-shaped IUD can be formulated to allow the escape of a known quantity-- usually the minute dose of 100 micrograms--of progesterone per day. Pharriss states that the uterine lining rapidly destroys natural progesterone so that essentially none of the steroid should migrate to other areas of the body. This should eliminate the side effects associated with other routes of administration. The mechanism of action of the device is unknown, but Pharriss thinks that it either prevents fertilization by preventing sperm capacitation (a maturation process required before the sperm are capable of fertilization) in the female reproductive tract or that it alters the uterine lining so that implantation cannot occur. The device that Alza is testing in this country is designed to protect against pregnancy for 1 year, but protection for 2 or 3 years is a feasible goal.

The "morning-after pill" has engendered widespread interest, especially since the revelation that the controversial synthetic estrogen, diethyl stilbestrol (DES) has been used rather routinely for this purpose. Although DES has been linked to the occurrence of a rare type of vaginal cancer in the daughters of women who took the drug during pregnancy to prevent miscarriages, the FDA recently approved the use of DES as a postcoital contraceptive in "emergencies"--after rape and incest--but has not approved it for routine use.

The development of a reliable, safe "morning-after pill" would be a major breakthrough in birth control technology. It would eliminate the need for long-term exposure to steroids and other potent drugs, particularly for women who have intercourse infrequently. The Contraceptive Development Branch of the National Institute of Child Health and Human Development (NICHD), Bethesda, Maryland, issued a Request for Proposals (RFP) for the study of estrogenic "morning-after pills." The RFP specifically excluded DES from consideration. The response to the RFP was low and it does not appear that enough women for satisfactory drug evaluation will be included in the funded proposals.

Estrogens are effective as postcoital contraceptives presumably because they speed the passage of the egg through the oviduct so that it arrives in the uterus before the lining is prepared for implantation of a fertilized egg. Several investigators pointed out that the relatively large doses of estrogens required to produce this effect may also produce nausea and other unpleasant symptoms and may thus make them unsuitable for routine use.

Progesterone, secreted by a structure called the corpus luteum, is required to prepare the uterus for implantation and also to maintain pregnancy during the first few months. (After the ovarian follicle releases the egg, the follicle is converted to the corpus luteum.) Therefore, luteolytic agents--materials that destroy the corpus luteum--will prevent or terminate pregnancy. Several unrelated chemicals including oxymetholone (a steroid), aminoglutathimide, and prostaglandins are being examined for their luteolytic capacities.

Prostaglandins have frequently been characterized as "miracle drugs"--and perhaps they are--but at present they are something less than miraculous as contraceptives. Several investigators have indicated varying degrees of disillusionment; Alza Corporation, for example, is scaling down, although not eliminating, its research program on the contraceptive action of prostaglandins. Nevertheless, Pharriss has found that some prostaglandins have luteolytic activity in subhuman species. Thus they have a potential use as postcoital contraceptives if similar activity occurs in the human.

According to Earl S. Gerard of the Upjohn Company, a luteolytic agent could be employed either to induce menstruation on schedule, whether or not conception had occurred, or to induce a delayed menstrual period. Prostaglandins do induce bleeding but the dose required also produces considerable discomfort, including nausea, vomiting, diarrhea, and cramps. Moreover, the bleeding may be the result of uterine contractions rather than luteolysis. Prostaglandins stimulate the contractions of the smooth muscle of the gastrointestinal tract (and thus produce the side effects) and of the uterus. For this reason they can be used to induce labor at term and as abortifacients.

PROSTAGLANDINS AND ABORTION

Prostaglandins are being used in Europe to induce abortions. Pharriss thinks that prostaglandins, probably PGF_{2a} and PGE_2, may eventually replace saline infusion as the method of choice for induction of abortion during the second trimester of pregnancy. In order to induce abortion, the prostaglandin may either be administered intravenously or it may be injected through the vagina into the uterus between the fetal membrane and the uterine lining. The second method produces fewer unpleasant side effects.

Most of the natural prostaglandins display a wide spectrum of systemic effects. The synthesis of prostaglandin analogs that possess only specific effects is the goal of investigators in several laboratories. Josef Fried of the University of Chicago has synthesized two such compounds; they have luteolytic activity, at least in animals, but only negligible activity in stimulating the smooth muscles of the intestinal tract. They are also resistant to the enzymes that normally deactivate natural prostaglandins. Although these results are encouraging, the effectiveness of the analogs in the human remains to be demonstrated.

Sterilization is, of course, a very effective means of preventing conception. Current sterilization procedures suffer from two major disadvantages: They are essentially irreversible, and they require surgery--relatively minor in the male and somewhat more serious in the female. Investigators at the Illinois Institute of Technology Research Institute (IITRI), Chicago, are developing improved techniques for the sterilization of both men and women. Erich Brueschke and his colleague, Marvin Burns, are using dogs to test a valve that would allow reversible sterilization. The valve is inserted into the sperm duct; when closed the valve would block sperm passage, but it could be opened later if desired.

Procedures for female sterilization that do not require abdominal surgery and that can be performed with local anesthesia and without hospitalization are also in the offing. Both Brueschke and Ralph Richart and Robert Neuwirth of the International Institute of Human Reproduction, Columbia University, New York, are developing instruments called hysteroscopes. These instruments, which are inserted through the vagina into the uterus, enable the physician to see the oviducts and to occlude them by an appropriate method. The investigators at Columbia have already used their hysteroscope to sterilize about 90 women, usually by electric cautery. The IITRI group is investigating both chemical and mechanical (use of plastic plugs) methods of occluding the tubes. They are testing the device and the sterilization techniques on baboons but are not yet ready to use them in the human.

The discovery of hormones secreted by the brain that regulate reproductive processes has opened a new and highly promising approach to the control of human fertility. This avenue of investigation is being explored in the laboratories of Roger Guillemin at the Salk Institute, La Jolla, California, and of Andrew Schally at the Veterans Administration Hospital, New Orleans, Louisiana. The hormone of interest--luteinizing hormone releasing hormone (LH-RH)--is secreted by a part of the brain called the hypothalamus and stimulates the release of luteinizing hormone by the pituitary gland. Luteinizing hormone in turn is the trigger for ovulation.

The releasing hormone is a relatively simple molecule--a peptide consisting of ten amino acids. Consequently, not only has the synthesis of LH-RH been accomplished, but also the synthesis of closely related chemical analogs. Both

Schally and Guillemin would like to synthesize analogs that block the activity of the natural hormone on the pituitary gland and thus prevent ovulation. Although Guillemin has reported synthesis of analogs with limited inhibitory powers, none have yet been found that are suitable for clinical trials as birth control agents. Both investigators are optimistic that such compounds will be synthesized in the future.

Use of LH-RH may also greatly increase the reliability of the rhythm method of birth control. Rhythm, the only method sanctioned by the hierarchy of the Catholic Church, frequently fails--about 25 percent of the time--partly because of inadequate knowledge of the time of ovulation. Administration of a suitable preparation of LH-RH may enable a woman to control precisely the time of her ovulation. According to Guillemin, an oral preparation is feasible, even though the peptide is susceptible to digestion in the gastrointestinal tract. The releasing factor is so extremely potent that if the oral dose is large enough, a quantity of LH-RH sufficient to cause ovulation should enter the blood stream. Finally, because it can induce ovulation, LH-RH may be employed in the treatment of some forms of infertility.

IMMUNOLOGICAL BIRTH CONTROL

The application of the immune system to reduce fertility is not an immediate prospect, but several investigators think that this approach may be possible in the future. Antigens specific to sperm have been identified, isolated, and used to immunize both males and females against the sperm that carry the antigen. For example, Erwin Goldberg at Northwestern University, Chicago, Illinois, found that serum containing antibodies to the sperm-specific form of the enzyme lactate dehydrogenase (LDH-X) suppressed the pregnancies of up to 60 percent of mice injected with the antiserum after copulation. The amount of suppression declined as the time between copulation and administration of the antiserum increased. Goldberg detected no pathological changes in the animals that had received antiserum. Moreover, the effect was reversible; when a group of the treated mice were later mated, all delivered normal litters.

The results of preliminary experiments--and Goldberg stresses the word "preliminary"--have encouraged him to think that the immune system can serve as the foundation of one type of male contraception. When Goldberg injected mouse LDH-X antibody into male rabbits, the fertility of the rabbits decreased in proportion to the concentration of LDH-X antibody in their blood. The effect was reversible, but a booster shot of the enzyme restored the infertility. Also encouraging was the observation that the rabbits did not suffer from diminished libido.

Oral contraceptives for males, although long predicted, remain elusive. Male fertility can be chemically depressed by preventing sperm formation entirely or by inhibiting sperm maturation. Several laboratories have programs to test various drugs for these activities in experimental animals, but few promising chemicals have been found as yet. Most investigators expressed the opinion that women would continue to bear the responsibility for contraception--as well as the consequences of its failure--for the foreseeable future. However, Alvin Paulsen and his colleagues at the University of Washington Medical School, Seattle, have been testing a potential male oral contraceptive in approximately 50 human volunteers. In early trials with Danazol, a synthetic analog of a male hormone, they did not observe consistent reduction in sperm counts; moreover, the concentrations of male sex hormones declined, as did libido in volunteers--a common problem when male hormone production is disrupted. In more recent experiments, Paulsen combined Danazol with a synthetic androgen and achieved more satisfactory results. Sperm counts were reduced without apparent side effects. Paulsen, citing FDA regulations for new drug development as his reason, did not care to estimate when full clinical trials of the drug would begin.

The declining birth rate in the United States has indicated the acceptance--at least for the present--of fertility control, as well as the methods available for achieving such control. Nevertheless no method is perfect for all situations. The goal of current research is to improve older technologies and to develop new ones so that the demands of diverse social, economic, cultural, and religious conditions can be met.

THE PANACEA PILL

In an event that one day promises to be ranked in medical history alongside the introduction of antibiotics and steroid drugs, the Upjohn Company not long ago put the first prostaglandins on sale in England, almost forty years after they were discovered. Two synthetic varieties, Prostin E_2, and Prostin F_2-alpha, were cleared for marketing in the United Kingdom, a situation that is likely to occur in many other countries before the drugs make their way through the regulatory undergrowth of our Food and Drug Administration. These first two commercially available prostaglandins are being used for the termination of early pregnancy and the induction of labor at term. But waiting not far behind are twelve more prostaglandins in this family of remarkable, naturally occurring substances that regulate many of the body's activities. Hormonelike in their action, prostaglandins occur in minute amounts in the tissues of mammals and some lower animals. As drugs, they are expected to be the near-perfect birth-control pill, as well as the medical answer to treating ulcers, controlling inflammation and blood pressure, and relieving asthma and nasal congestion. All before the end of this decade.

Copyright 1973 by Saturday Review Co. Used with permission.

BREAKFAST OF CHAMPIONS
Kurt Vonnegut, Jr.

Human mental activity embodies an incredibly complex interplay of chemicals, and it's not surprising that psychiatrists are finding that mental disorders can arise from biochemical aberrancies. Linus Pauling is looking for cures to mental illness through massive doses of vitamins. As Kurt Vonnegut and R. D. Laing show, however, there are plenty of causes for insanity within the dynamics of human relations. To the extent that insanity is a social phonomenon, it would seem that the cure must lie in social change.

Everybody in America was supposed to grab whatever he could and hold onto it. Some Americans were very good at grabbing and holding, were fabulously well-to-do. Others couldn't get their hands on doodley-squat.

Dwayne Hoover was fabulously well-to-do when he met Kilgore Trout. A man whispered those exact words to a friend one morning as Dwayne walked by: "Fabulously well-to-do."

And here's how much of the planet Kilgore Trout owned in those days: doodley-squat.

And Kilgore Trout and Dwayne Hoover met in Midland City, which was Dwayne's home town, during an Arts Festival there in autumn of 1972.

As has already been said: Dwayne was a Pontiac dealer who was going insane.

Dwayne's incipient insanity was mainly a matter of chemicals, of course. Dwayne Hoover's body was manufacturing certain chemicals which unbalanced his

From *Breakfast of Champions: Or Goodbye Blue Monday*. Copyright © 1973 by Kurt Vonnegut, Jr. A Seymour Lawrence Book/Delacorte Press. Reprinted by permission of the publisher.

mind. But Dwayne, like all novice lunatics, needed some bad ideas, too, so that his craziness could have shape and direction.

Bad chemicals and bad ideas were the Yin and Yang of madness. Yin and Yang were Chinese symbols of harmony. They looked like this:

The bad ideas were delivered to Dwayne by Kilgore Trout. Trout considered himself not only harmless but invisible. The world had paid so little attention to him that he supposed he was dead.

He hoped he was dead.

But he learned from his encounter with Dwayne that he was alive enough to give a fellow human being ideas which would turn him into a monster.

Here was the core of the bad ideas which Trout gave to Dwayne: Everybody on Earth was a robot, with one exception--Dwayne Hoover.

Of all the creatures in the Universe, only Dwayne was thinking and feeling and worrying and planning and so on. Nobody else knew what pain was. Nobody else had any choice to make. Everybody else was a fully automatic machine, whose purpose was to stimulate Dwayne. Dwayne was a new type of creature being tested by the Creator of the Universe.

Only Dwayne Hoover had free will.

Trout did not expect to be believed. He put the bad ideas into a science-fiction novel, and that was where Dwayne found them. The book wasn't addressed to Dwayne alone. Trout had never heard of Dwayne when he wrote it. It was addressed to anybody who happened to open it up. It said to simply anybody, in effect, "Hey--guess what: You're the only creature with free will. How does that make you feel?" And so on.

It was a tour de force. It was a jeu d'esprit.

But it was mind poison to Dwayne.

WILL VITAMINS REPLACE THE PSYCHIATRIST'S COUCH?
Robert J. Trotter

"I told you so," says Linus Pauling about vitamin C and the common cold. And he hopes to be saying the same thing about vitamin B_3 and mental illness. But the way things stand now, the two-time Nobel Prize winning chemist may have to wait a while before he takes his bows.

Orthomolecular psychiatry, the term Pauling coined for the treatment of mental illness with massive doses of vitamins, is meeting strong resistance from establishment psychiatry. Only two percent of the nation's psychiatrists, estimates Pauling, use orthomolecular therapy. The National Institute of Mental Health has discouraged such treatment. And the American Psychiatric Association is preparing to release a task force report that says the theoretical basis for megavitamin therapy has been "found wanting" and attempts to prove its value have been "uniformly negative."

This reaction is natural. Since the time of Sigmund Freud, psychiatry has come to rely heavily on the psychological model of mental illness as an immaterial disorder or perversion of the intellect. But "psychotherapy has failed," says Pauling, and it is time to get back to the medical or scientific model. One way to do so, he feels, is with vitamins.

The relation of vitamins to mental illness has been evident since vitamins were discovered. Pellagra, for instance, is a vitamin-deficiency disease, and one symptom of it is psychosis. In 1937 researchers discovered that an adequate intake of niacin (a B-complex vitamin) could avert or cure both the disease and the psychosis. Similar treatment was applied to other forms of mental illness with varying amounts of success in the 1940's. Then in 1952, Humphry Osmond and A.

Reprinted by permission of *Science News*, © 1973 by Science Service.

Hoffer began giving niacin to patients diagnosed as schizophrenic. Their published results claimed significant improvement in these patients when compared to patients receiving a placebo. Even with these results, megavitamin therapy received little serious attention until 1968 when Pauling lent his support to the theory.

In an article in Science Pauling hypothesized "that the so-called gene for schizophrenia may itself be a gene that leads to a localized cerebral deficiency in one or more vital substances." Orthomolecular psychiatric therapy, he said, attempts to overcome this deficiency by providing "the optimum molecular environment for the mind, especially the optimum concentrations of substances normally present in the human body." Pauling cited the work of Osmond and Hoffer and then noted that mental illness can result from a low concentration in the brain of any of the following vitamins: thiamin (B_1), nicotinic acid or nicotinamide (B_3), pyridoxin (B_6), cyanocobalamin (B_{12}), biotin (H), ascorbic acid (C), and folic acid. Replacing shortages of these vitamins by orthomolecular or megavitamin therapy, he said, "may turn out to be the best method of treatment for many patients."

Early this year Pauling and David Hawkins of the North Nassau Mental Health Center in New York co-edited and published a volume titled Orthomolecular Psychiatry. In it, 37 articles by 30 authors set forth the theoretical, experimental and clinical background of orthomolecular psychiatry and how it relates to schizophrenia, alcoholism and mental disorders resulting from drug abuse. Hawkins describes how the orthomolecular approach can be applied inexpensively to large numbers of patients in a clinic setting. The book also contains a description of a self-help organization known as Schizophrenics Anonymous.

It was this kind of publicity that helped get megavitamins off the shelf and into the headlines. Megavitamin therapy has become something of a fad in various parts of the country. But according to NIMH and the APA, it is not what the doctor called for. The most recent issue of NIMH's Schizophrenia Bulletin says the bulk of research evidence shows that megavitamin therapy adds nothing to the usual psychiatric treatments.

The APA comes down a little harder. The task force on vitamin therapy in psychiatry says the results and claims of megavitamin therapy have not been confirmed. It cites the work of several groups of psychiatrists and psychologists who do psychopharmacological research. Says the task force: "It is virtually impossible to replicate studies in which each patient receives a highly individualized therapeutic program with from one to seven vitamins in huge doses, plus hormones, special diets, other drugs and electroconvulsive therapy, which are added or subtracted not on the basis of proved biochemical abnormalities but rather on the basis of the clinicians' individual judgment as to the patient's need... It is also impossible to replicate studies in which as many as five years of treatment may be needed before results begin to appear." About orthomolecular therapy's reported successes, the task force says, "one must seriously doubt that the patients were all truly schizophrenic."

Possible long-range toxicity is considered by the task force to be "the truly important question arising from the prolonged megadosage administration of B_3." Toxic reactions in humans, the report says, include duodenal ulcer, abnormal liver function, hyperglycemia and extraordinary increases of serum uric acid. Another concern of the report is that mental health clinics may be put out of business by the creation of new orthomolecular clinics that may not offer the scope of service given by more conventional clinics. And finally, the task force finds "deplorable" the massive publicity that megavitamin proponents promulgate via radio, the lay press and popular books.

"Deplorable from whose standpoint?" asks Pauling, who says he is happy to be making a contribution. "These are the same statements physicians made about what

what I said about vitamin C and the common cold. And it's quite clear," he told Science News, "that the physicians were wrong. All the new evidence that has come out has shown that the statements I made were correct."

(Pauling is correct in saying that some research has come out in his favor. Studies in Czechoslovakia have found that vitamin C helps remove cholesterol from the body and therefore protects against atherosclerosis and heart attack [SN: 2/17/73, p. 106]. Studies with college students in Canada have shown that megadoses of vitamin C can protect against the common cold. Studies at the University of Texas suggest that even more vitamin C than Pauling calls for is necessary for good health and normal development [SN: 5/5/73, p. 290]. But this evidence does not convince everyone. Nutritionists still call for only one-tenth of what Pauling suggests. Pauling, however, is not talking about nutrition. Megavitamins, he says, have a therapeutic and pharmacological effect. Paul E. Johnson of the Food and Nutrition Board of the National Academy of Sciences says no one really knows what will finally come of the therapeutic and pharmacological use of megadoses of vitamin C. What is needed, he says, is an unbiased review of all the research on the subject. This has not been done.)

Pauling says more than 20,000 patients have received up to 8 grams of niacin and 4 grams of ascorbic acid a day, and there has been no evidence of long-term toxicity or the side effects mentioned by the APA task force. Pauling admits that diagnosing schizophrenia isn't easy but, "We accepted as schizophrenic patients who were diagnosed so by two psychiatrists."

Orthomolecular psychiatry is not aimed at destroying psychiatry or mental health clinics, Pauling emphasizes. Instead, megavitamins, are supposed to be auxiliary to the other therapeutic and prophylactic measures used by psychiatrists. They supplement, not replace, other therapeutic treatments. Why then does psychiatry seem reluctant to accept orthomolecular therapy? The old guard doesn't want to learn anything new, Pauling says, and "if the patients go back to a normal life faster, then the number of patients the psychiatrists have to treat will be less and there may be a danger to their income right there."

*

>   There must be something the matter with him
>   because he would not be acting as he does
>   unless there was
>   therefore he is acting as he is
>   because there is something the matter with him

He does not think there is anything the matter with him
because
       one of the things that is
       the matter with him
       is that he does not think that there is anything
       the matter with him
therefore
       we have to help him realize that,
       the fact that he does not think there is anything
       the matter with him
       is one of the things that is
       the matter with him

From *Knots* by R.D. Laing. Copyright © 1970 by The R. D. Laing Trust. Reprinted by permission of Pantheon Books/ A Division of Random House, Inc.

I don't feel good
therefore I am bad
therefore no one loves me.

I feel good
therefore I am good
therefore everyone loves me.

I am good
You do not love me
therefore you are bad. So I do not love you.

I am good
You love me
therefore you are good. So I love you.

I am bad
You love me
therefore you are bad.

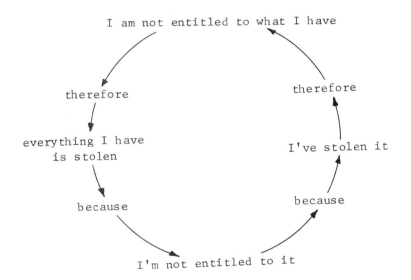

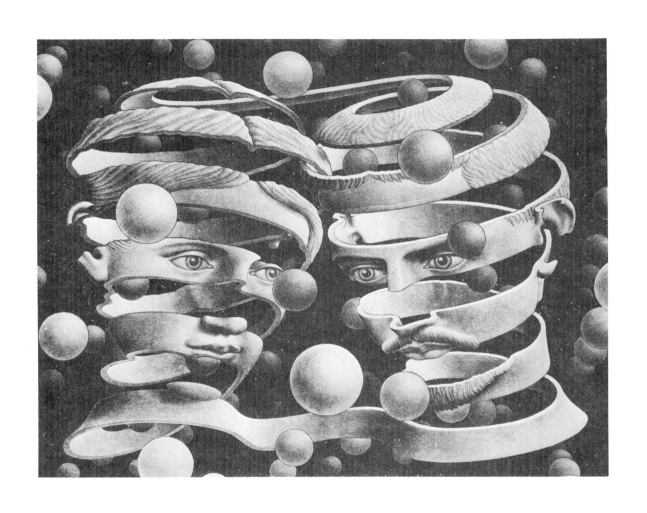

PART V
GENETICS

CREEPING UP ON EUGENICS
Phillip Chapnick

A technology for the manipulation of human genes is
developing. The question is: How will it be used?

New York, March 18 - A genetic screening program to test for two genetic diseases among the Jewish population, beginning with residents of Westchester County and other areas of metropolitan New York, was announced today by the Albert Einstein College of Medicine and the National Genetics Foundation.

Tay-Sachs Disease is a recessive hereditary illness which strikes infants and causes death at three or four years of age; hyperlipidemia, a dominant hereditary disorder of fat metabolism, may cause a heart attack at a relatively early age. The genetic screening program for these diseases is simple: a blood test reveals whether the individual is a carrier of Tay-Sachs or suffers from hyperlipidemia. Couples tested learn their chances of having a child with these disorders; if they choose to have children, amniocentesis tests which sample fluid surrounding the fetus can determine whether it is afflicted in utero. If so, the parents may opt for a legal abortion on medical grounds.

The Einstein-NGF program appears to be a harbinger of much more widespread genetic screenings; similar programs are envisaged for such diseases as sickle cell anemia which is found mainly in blacks, cystic fibrosis which is found in whites, phenylketonuria which strikes persons of Irish or North European ancestry, and thalassemia major which afflicts Italians and Greeks. "It is better to

From *The Sciences*, May 1973. Copyright © 1973 by The New York Academy of Sciences. Used with permission.

recognize a condition before it takes its toll on health," says Dr. Arthur Steinberg, Chairman of the Medical Advisory Board of NGF.

DECREASING THE BAD GENES

Although proponents emphasize health, screening programs will have the same effect as "negative" eugenics programs would: reducing selected deleterious genes from the nation's gene pool to the level of such genes introduced by spontaneous mutations. The list of diseases linked to a single dominant gene is staggering--and growing. They number about 415, and affect perhaps one per cent of the population--or more than two million people. For their elimination, the technique of choice is abortion.

Huntington's chorea--a progressively degenerative disease affecting involuntary movements and mental functioning--does not usually appear in individuals until after their procreative years, and is thus not selected against by nature. Carried by a dominant gene, it would be one of the best targets for a negative eugenics program--if an accurate method of determining its presence in utero were developed. Drs. Arno Matulsky, George Fraser and Joseph Felsenstein, all of the University of Washington School of Medicine, have calculated the number of abortions required to reduce the frequency of the Huntington's chorea gene to that of spontaneous mutation. Assuming that every couple has the goal of having two normal children, they estimate that about 2,000 abortions would be required for every 500,000 couples; if the entire population were screened, the disease could be nearly eliminated.

THE STRENGTH OF RECESSIVE DISEASE

Diseases produced by recessive genes--about 365 are known for certain--apparently could not be eliminated by abortion. "If you're talking about situations [involving recessive genes] where the abnormality develops only in those who have the double dose, the problem is that most of the genes you would like to be eliminating exist in the population in single dose," Dr. William Schull, Director of the Center for Population and Demographic Genetics, University of Houston, said in a telephone interview. "You could remove the affected [double-dosed] persons from the gene pool, but this is like chipping off the top of the iceberg."

Matulsky et al calculate that reduction of the frequency of the recessive gene for cystic fibrosis to its spontaneous mutation level would require 100,000 abortions per 500,000 couples; similar reduction of the recessive sickle cell gene would require twice as many abortions. They conclude that "selective abortion is not an appropriate method for the elimination of the genes causing sickle cell and cystic fibrosis since a forbiddingly large number of heterozygote clinically normal fetuses would be required." (Birth Defects: Original Article Series)

Fears have been expressed that selective abortion of the fetuses of parents with a double-dose of a recessive gene may actually have dysgenic effects. Selective abortion may serve to increase the frequencies of bad genes in the population because parents could choose to have clinically normal children, and a high proportion of these offspring would be carriers. Matulsky and his co-authors say that such dysgenic fears are unfounded. "The potential dysgenic effects of selective abortion (as well as of gamete selection or mating control) are so minor as to be trivial for a number of generations," they write, "which is well beyond any reasonable projection of the need for employment of such programs."

IMPROVING THE SPECIES?

Positive--as opposed to negative--eugenists aspire to loftier goals: the genetic improvement of the human race. No such programs are underway, but a very well known positive eugenic proposal was put forward by the late Nobelist H. J. Muller. Sperm from specially selected donors would be stored until dispassionate judgment could be made of the donors' accomplishments in life. The sperm of donors judged outstanding would be used for artificial insemination, at first only in cases where husbands are infertile, eventually perhaps for any woman who chose to avail herself of the service.

Along this line are thoughts suggested by recent advances in cell cloning techniques. It is now possible to plant the nucleus of a living cell into a fertilized egg (whose nucleus has been removed), thus transferring en masse the cell donors' entire genetic constitution. If this technique could be applied to human beings, and if the "new" fertilized egg could be nurtured in a real or artificial womb, the infant which resulted would be an exact biological twin of the cell donor when he or she was newborn.

Questions of their scientific possibility and social acceptability aside, positive eugenic programs might be genetically disadvantageous. One of nature's ways of continuing evolution is to provide genetic variability within a species; sensitive gene typing techniques have revealed a great deal more human variability at the gene level than was previously suspected. Would this vital variability be reduced if a small group of selected genes were specially propagated? This question probably could not be answered until after the fact.

FROM BAD TO GOOD SCIENCE

Francis Galton, a cousin of Charles Darwin, is considered to be the founder of the modern eugenics movement. To decrease human suffering and increase the general welfare, Galton suggested that successful people be encouraged to procreate more, "unsuccessful" ones to procreate less. This neglect of environmental factors led to a spate of unscientific works, which were influential in providing support for biased immigration laws passed in the U.S. in the early 1900s. Also based on bad science, Hitler's genocidal attempt to eliminate the "inferior" peoples of Europe and breed a pure master race strengthened the fear many people have of the eugenics movement in social policy decision-making.

Recent eugenic proposals, to their credit, are dependent upon good science: the basic study of population genetics. Every human cell contains more than ten thousand genes; the closer that each bit of genetic information fits the environment, the more likely it will stay in the collective genetic bank of our species. As Dr. James Crow, Bascom Professor of Genetics at the University of Wisconsin at Madison explains it, "Selection is simply this: one group of plants or animals leaving a larger group of descendants than another. In the long run that changes the gene frequencies; and in the longer run that changes the average [value of some measured characteristic]."

A mathematical theory of evolution has been developed by applying the theory of probability to these Mendelian inheritance principles. The theory looks on a population as a collection of genes, specified at any given time by the number and proportions of different alleles. Population geneticists seek to determine and predict how gene frequencies change from generation to generation by constructing more detailed mathematical models, including such factors as a spontaneous mutation rate, relative genotypic viability and polygenic inheritance.

GENES AND SOCIETY

Anything that affects reproduction rates--predators, availability of food, climate, geography--can have evolutionary consequences. Perhaps the most important factors for <u>Homo sapiens</u> are of social origin: social structures determine who marries and how many children are produced, medical advances enable people to procreate who otherwise would have died without leaving children, civilization exposes us to increasing doses of mutagenic radiation and chemicals. (Surveys on radiologists and others who come into frequent contact with radiation show higher rates of spontaneous abortion and abnormalities among their offspring; present estimates are that a single 50-to-150 roentgens exposure doubles the individual's genetic mutation rate. "The business of chemical mutagenesis is much more difficult to evaluate," Dr. Schull says. "Before inhaled or ingested chemicals get to [the gametic cells], they have to pass through lots of filters; even if mutagenic <u>in vitro</u>, they may not be so <u>in vino</u>. But most of us feel that chemicals have a more important mutagenic effect than radiation.")

GENETIC OVERLOAD?

One reason why eugenics is being reconsidered even after its dubious history: the presumption that social and environmental factors are increasing the number of mutations in the population. H. J. Muller, who discovered that radiation increases the mutation rate of drosophila, introduced the term "genetic load"--an increasing accumulation of mutations which lessens a population's ability to survive and reproduce. A subtle increase in load could have significant long-range consequences. "If a mutation causes very mild effects, say a two or three per cent reduction in survival probability," says Dr. Crow, "eventually it will make an impact on the population."

However, he points out that genetic load is very hard to measure in human beings. "The effect of an increased mutation rate could simply be an increase in the incidence of things we already have. If we have a ten per cent increase in some kind of abnormality, we probably wouldn't notice it without very careful records. Even then, we couldn't be sure it wasn't due to some other cause than mutation."

GENETIC RELATIVISM

According to Dr. Schull, even a heavy genetic load may <u>not</u> mean general human deterioration. "To assume that a population is deteriorating seems to imply some fixed set of environmental circumstances: that is, you are comparing what's happening to the biology of a population to some benchmark. If that benchmark keeps changing with time--if the genetic needs of the population are altering from time to time--it is difficult to see how one measures either progress or decay."

Changing environmental conditions must be taken into account in evaluating a particular genic constitution, says Dr. Schull; genes may be advantageous or disadvantageous under different conditions. "There is some evidence to suggest that a diabetic individual, if placed in a situation where his food intake were marginal, would be at a relative advantage to the remainder of the population. His overly efficient metabolic system would extract relatively more from what he ingests than would those of average individuals." Another example: the gene for sickle cell anemia provides protection against malaria in single dose and is selectively favored in areas where malaria is endemic.

ANY POINT TO EUGENICS?

Social determinism is a frame of thought which suggests that man can be understood primarily as a social organism: social, political and cultural factors are the critical ones. Biological determinists, on the other hand, hold that genes are most important variables in understanding behavior. "The Fallacies of Biological Determinism" was the title of a recent meeting of the Science and Social Policy Section of the New York Academy of Sciences. The speaker: Dr. Richard Lewontin, an eminent population geneticist soon to be affiliated with Harvard University, and--not surprisingly--an avowed social determinist.

Lewontin chided those who seek human improvement through eugenics. Comparing the slow rate at which any genetic changes could be propagated through the population to the incredibly rapid pace of technological and social changes, he said that eugenic plans were trivial.

Whether people like Dr. Crow would agree is very doubtful. Admittedly more of a eugenist than many of this colleagues, Dr. Crow thinks that directed genetic selection can be effective. "My view is that we can change the population a great deal in various directions--if we could agree as to how we want to do it," he told me. "My main reason for not becoming an active eugenist is that I have social reservations--fear of misuse and a belief that society isn't quite ready for this type of thing.

"But," Dr. Crow added, "I do want to start to discuss it."

*

PRECAUTIONS FOR PROSPECTIVE PARENTS

It will be some time, more like decades than months, before science has arrived at the stage where it is able to cure most birth defects and prevent those that it cannot cure. But there are a number of common-sense precautions every prospective parent should know.

1) A hereditary condition in the family means that one should get some genetic counseling.

2) It is best for a woman to have her children between the ages of twenty and thirty-five, with at least two years between the end of one pregnancy and the beginning of another; the father should be younger than forty-five years old.

3) Intercourse should take place at intervals of no more than twenty-four hours for several days just preceeding and during the estimated time of ovulation.

4) No woman should become pregnant unless she has had German measles or has been effectively immunized against it, and during pregnancy she should do everything possible to avoid exposure to contagious disease.

5) A woman should avoid eating undercooked red meat or contact with cats that have eaten such meat or that are allowed to hunt, since either may be the source of an organism known as toxoplasma that causes major birth defects.

6) A woman who is pregnant or who thinks she is should not be exposed to X rays or take any drugs (except under the instructions of a doctor who has been told of her pregnancy).

There are several other items, including good prenatal care, not smoking cigarettes, and getting appropriate treatment if she is Rh negative, that most obstetricians tell their patients once pregnancy has been established. Practicing

Copyright © 1973 by Saturday Review Co. Used with permission.

these guidelines won't guarantee a healthy, normal baby, advises pediatrics professor Virginia Apgar in *Is My Baby All Right?* published earlier this year. But they will greatly increase a couple's chances of having children free of congenital handicap.

DOOMED TO INEQUALITY?
Gerald Chasin

Arthur Jensen shook the country with the publication of his theory that the black race is less educable then the white. In this review of his recently published, <u>Genetics and Education</u>, an unbiased summary of Jensen's work is attempted, and the major arguments against Jensen's theories are summarized.

In the winter of 1969 Arthur Jensen published an article in the <u>Harvard Education Review</u> which approached monographic proportions in length. It was the longest article ever to appear in that journal and stirred up one of the most heated controversies in the social sciences. In it Jensen claimed to have amassed proof that the average difference between blacks and whites of 15 points in intelligence test scores is due to differential genetic inheritance in the two races. In other words, the 15-point deficit which blacks show in comparison to whites is due to their genetic inferiority to whites. This thesis immediately provoked a barrage of protests and technical refutations. The <u>Harvard Education Review</u> refused to print it without including a series of critical articles, and followed in its next issue with another set of refutations. The volume which Jensen has now put out consists of his original <u>Harvard Education Review</u> article along with three other pieces, two having to do with the inheritance of intelligence, and the other dealing with mental retardation. All of these papers have been published elsewhere in easily obtainable sources. The only new writing which this volume contains is a lengthy 67-page preface in which the author attempts to make his position clear and to explain to the public that he has been unfairly treated

Reprinted from *The Nation*, July 2, 1973. Used with permission.

by both professional and nonprofessional audiences.

In his preface Jensen claims that the protests and criticisms his article drew were in large part politically and ideologically motivated, rather than being objective scientific statements. The author portrays himself as a sober, rational and diligent scientist who has always been fully committed to the canons of free scientific inquiry. His opponents, by contrast, have been biased in not accepting the scientific proof he offered in any open-minded manner and have been swayed by their social values, not by critical reasoning. Jensen also complains of having been thoroughly harassed, being the object of large amounts of hate mail, threatening telephone calls and intrusions into his lectures. He accuses the press of misinforming the public about his views and of casting him in the role of villain. Jensen reports that he has been demonstrated against, been forced to lecture in secret, and has required the company of plainclothes men for personal protection around the Berkeley campus, all as a consequence of his honesty and forthrightness in stating the truth about an unpopular issue.

According to Jensen, not only have students and the press treated him in a grossly unfair way, but his professional colleagues and the professional establishment have also reacted toward him in an unethical and narrow-minded fashion. He claims that many of his academic opponents have criticized his article without first having read it, and the professional journals barred the publication of further articles on the genetic inheritance of intelligence. Jensen also asserts that the Harvard Education Review, which at first had solicited his article, later disclaimed having done so once the protests began. On the whole, Jensen feels that the articles published in the journal did not amount to serious criticism and were full of ad hominem attacks. He says that these critiques also contain a large number of theoretical, factual and methodological errors which he has not been given the opportunity to refute. Because of the attitude of the Harvard Education Review and the professional associations, the author maintains, the principle of free discussion and freedom of inquiry has suffered a grave setback.

The preface to this volume contains many testimonials to the validity and significance of the author's ideas: statements by professional colleagues supporting him--especially geneticists; college courses and seminars in education, psychology and genetics which have discussed his article are mentioned. And Jensen is still convinced, on the basis of "massive evidence," that the educational aptitudes and needs of the majority of white and black children are sufficiently different that a diversity of educational aims and approaches is necessary and best for both groups, and particularly the more disadvantaged one. The effort we have put into equalizing scholastic performance has been misdirected, he believes. In his view more effort should be put into constructively diversifying schools. Since, as he claims to have shown, blacks have difficulty with conceptual learning and do better at rote learning, the schools should somehow reflect this racial difference. What this would mean, in effect, is that blacks would be sent to trade schools while whites would go into academic programs leading to better paid and more prestigious jobs.

Aside from the preface, the heart of the present volume is Jensen's original Harvard Education Review article, "How Much Can We Boost IQ and Scholastic Achievements?" In this paper Jensen begins with the assertion that compensatory education programs have failed to produce any significant and lasting effects, though they have been practiced on a large scale in many cities for years. The reason for this is that genetic factors place a ceiling on scholastic achievement. Jensen believes that intelligence tests are valid measures of important mental abilities and provide an index of the ability to adapt to our system of formal education, and to the occupational system with which it is linked. In one place he talks of intelligence tests as measuring the ability to adapt to Western civilization.

In Jensen's view, though genetic factors have been belittled by environmentalists, they are nevertheless significant. Citing the results of selective breeding with animals, the author discusses what he feels are the likely genetic results of assortive mating—the tendency for individuals to marry those of like intellectual level. Because of assortive mating a population is bred whose various groups grow more and more differentiated in respect to mental ability. Assortive mating leads to a dysgenic trend in the black population. Since the lower classes are less intelligent (in the black as well as the white population), and are likely to transmit their lack of mental ability to their offspring through the inferior genes they carry, a higher birth rate in the lower classes of the black population would tend to lower the overall intelligence of blacks. Jensen feels that research should be directed to the educational and social implications of these trends, and implies that current welfare policies which are unaided by "eugenic foresight" could lead to the "genetic enslavement" of a substantial segment of our black population. This would, in the author's view, constitute a "grave injustice" to the American Negro.

A major portion of Jensen's original article is concerned with the degree to which intelligence is inherited. The evidence consists of studies of individuals who are related in varying degrees, and who could hence be presumed to be, in varying degrees, genetically similar. From these studies, especially studies of identical twins who have been reared apart (and who have thus been subjected to different environmental influences), Jensen concludes that the heritability of intelligence is high; approximately 0.80 of one's intelligence is inherited, the rest is produced by the effects of environment, and by a combination of environmental and genetic effects. Accordingly, the environment has a small effect on intelligence; at most it has a "threshold effect." That is, too poor an environment can prevent a person from achieving the level of intelligence that he has been genetically endowed with, but any further improvement in environmental stimulation has no additional effect on intelligence. Black children, according to the author, are not reared in such extreme cultural deprivation that they can be said to be below the threshold level, and their lack of ability must be attributed to genetic and not environmental causes.

Strictly environmental explanations of black-white differences are extremely implausible to Jensen—they are unscientific and have an *ad hoc* quality. In his estimation, blacks are simply genetically inferior to whites. Since races are breeding populations who, through geographic and social isolation for many generations have come to differ in their gene distribution, Jensen feels it is not unreasonable to conclude that the average difference of 15 IQ points between blacks and whites is due to differences in genetic inheritance.

To the uninitiated, Jensen's writing appears to glisten with technical expertise. Until publication of his *Harvard Education Review* article, the author would probably have been recognized as one of the leading men in the field of learning and intelligence testing. But Jensen, in his preface, has drawn an inaccurate picture of his scientific integrity and of his alleged mistreatment. The mass media have been highly favorable in their response, rather than negative, as the author claims. Within two weeks of the *Harvard Education Review* article, *U.S. News & World Report* published a feature story about Jensen and his work; it made headlines in the Virginia papers; it was written up in *Saturday Review*; and *The New York Times Magazine* published a highly favorable account in the summer of 1969. A very recent example of a positive comment in the press is the story about the present volume that appeared in Newsweek last March.

In addition, a Congressman has read Jensen's article into the Congressional Record and segregationists have cited it in court. Jensen's professional colleagues generally have found it unconvincing, however, despite its liberal use of experimental and statistical data. Within the scientific community, accordingly, Jensen is not accepted, and there is little danger within that community of the further spread of a false and socially pernicious doctrine. But outside the boundaries of the campus and the lecture hall the danger is great. In giving favorable publicity to Jensen's views, the popular press has been responsible for spreading a potentially harmful and dangerous theory that is directly akin to the racism, social Darwinism and eugenics thinking of the 19th century, and which was believed--until Jensen published his Harvard Education Review article--to have been completely discredited.

Professional observers have almost universally found Jensen's technical argument unacceptable. Critics have pointed out that observed differences of 15 IQ points between blacks and whites could easily be consistent with an identical distribution of genes in the two races. Others have stated that the evidence supporting the theory that intelligence is largely inherited is itself weak, being based on a limited number of studies of identical twins who have been reared apart. Such studies have been done only on American and Northern European populations and have used small samples which have yielded statistically unreliable results. There is nothing whatsoever known at present about the degree to which intelligence in nonwhite populations is inherited, and what is known about white populations is dubious.

Most psychologists consider the heritability of intelligence as being far lower than Jensen's estimate. Critics also point out that studies of the similarity of intelligence of identical twins, from which the degree of heritability of intelligence is deduced, contain many methodological difficulties and that environmental rather than genetic factors could easily explain their results. For example, it has been mentioned that the separate environments in which identical twins have been reared are often not that different. One of the classic studies which Jensen relies heavily upon, done in 1937, was reanalyzed and shown to produce conclusions which do not support the genetic thesis. In the new analysis it was found that those twin pairs reared in similar environments had similar IQ test scores, while those reared in dissimilar environments had quite different IQ test scores. Still other critics have pointed out that even if the heritability of intelligence were high within a population, that would still not permit the conclusion that differences in the two groups were due to genetic factors. They may be, or they may not be. There is simply no evidence for it. In the case of the mean IQ difference between blacks and whites, there is basically no good reason to believe that genetic factors are responsible; Jensen has not proved this and he cannot do so because the data simply do not support the conclusion.

More than one of the experts commenting on Jensen's work have claimed that he has used selective and sometimes inappropriate sources, that he offers no new information but presents only a reorganization of old data, and that there are substantial distortions in his summaries of research. Many of the sources which Jensen cites as sound have been criticized for being methodologically inadequate, such as the Coleman Report. Audrey Shuey, whose book, The Testing of Negro Intelligence, Jensen leans upon heavily, is a known exponent of white intellectual superiority. Jensen's claims to scientific objectivity are therefore suspect.

In attacking Jensen, anthropologists have argued that race is an inappropriate concept, particularly when applied to man. The African populations from which black Americans are descended, rather than being homogeneous, are diverse both ethnically and genetically. Geneticists, in addition, have pointed out that it is a misleading oversimplification to regard the present-day black population in the United States as genetically homogeneous itself.

Still other critics have mentioned that compensatory education has not really failed, for it has not really been tried; it has suffered from a paucity of funding which has not allowed the application of continuous, planned efforts. A further error which Jensen makes is in equating social classes across the racial and caste barrier (a black man is treated differently from a white man in our society no matter how much he earns), and that the attempt to control statistically the effects of social class on the IQ scores of blacks is ill founded.

Given the evidence which Jensen produces, his conclusions are unreasonable. When one considers the culture-biased nature of intelligence tests, which have been constructed to apply to a white, middle-class population, the additional negative effects of fear and unfamiliarity with the testing situation, of being taught by members of another race and of a higher social class, of the ghetto environment, and of discrimination at all social levels, it is ridiculous to argue that a mean difference in intelligence-test scores of 15 points should be attributed to differential genetic inheritance.

Gerald Chasin teaches sociology at the Polytechnic Institute of Brooklyn. He is the author, with Barbara H. Chasin, of a book to be issued this spring, <u>Power and Ideology: A Marxist Approach to Political Sociology</u> (Schenkman).

*

BOARD OF EDUCATION REJECTS CONCEPT OF RACIAL SUPERIORITY

Reacting to criticism from the American Civil Liberties Union (ACLU), the board of education adopted two resolutions rejecting theories of racial superiority.

Both motions were introduced by Board Member Julian Nava and passed easily despite criticism that they were "superfluous."

The motions stemmed from appearances by ACLU officials who accused the board of covertly harboring a belief of genetic superiority.

Marvin Schachter, president of the ACLU's local chapter, appeared at the board meeting last Thursday and said the board in a court case has stated a position of "agnosticism" on the question of racial superiority.

INFERIOR RACE

This, Schachter argued, means that the board has an "operative assumption that black are an innately inferior race." He called on the board to "repudiate" any theory of racial superiority.

ACLU Education Chairman John Caughey appeared at Monday's board meeting and urged the board "...reject the demeaning dogma of racial inferiority" for the sake of all Americans.

In arguing for his motions, Dr. Nava said passage was "important" so the board could "clear the air" on where it stands.

"It would be a valuable public service to make it clear what the position of this board is," he said.

Reprinted with permission from *Los Angeles Sentinel*, September 27, 1973.

Board members J. C. Chambers and Richard E. Ferraro argued against the motions, saying they were "superfluous" and "unnecessary." Ferraro, however, voted for both, while Chambers abstained on one.

Dr. Nava introduced the motions as one, but they were separated following a motion by Dr. Chambers.

The first motion passed 7-0. It reads:

> "The Los Angeles Unified School District reaffirms its dedication to the principle of equal educational opportunity for all students without regard to race, color, creed, sex or ethnicity, and that all actions and policies of the school district will reflect this principle."

The second motion passed 6-0-1, with Dr. Chambers abstaining, reads:

> <u>Create Equal</u> "The Board of Education of the Los Angeles Unified School District declares that it believes all mankind is created equal and it rejects as false all assertions that any race is inherently superior to another."

Board members debated that issue for more than an hour.

The issue was raised last Thursday when ACLU president Schachter quoted a brief filed by school board attorneys in the 1969 lawsuit.

Schachter quoted that brief as saying, "We do not assert that there are in fact distinctive ethnic-group patterns of abilities... nor do we assert there are not such patterns. We are agnostic in this regard."

A 126-UNIT ARTIFICIAL GENE

Some of the recent experiments in genetics are beautiful in their design.

One of the biggest challenges in molecular biology is determining the chemistry of genes and how this chemistry orders genetic information. A major step toward this end was announced in 1970 by Nobel laureate chemist Har Gobind Khorana and his team. They were the first scientists to synthesize a copy of a real gene. The gene, found in yeast cells, orders transfer RNA to line up the amino acid alanine into protein (SN: 6/6/70, p. 547).

The artificial gene had two drawbacks though. Because the start and stop signals for the alanine tRNA gene were not known, the scientists could not attach these signals to the artificial gene. And without the start and stop signals, the artificial gene could not be made to function in a yeast cell. And even if the artificial gene could have functioned in a yeast cell, alanine tRNA activity ordered by it could not have been detected. This is because the yeast cell's natural genes activate alanine tRNA as well.

Khorana and his team have now synthesized another gene. This one is the first synthesized gene that has potential for both functioning in a living cell and for having its product detected in a living cell. The achievement was reported this week by Kanhiya Lal Agarwal, one of Khorana's colleagues at the Massachusetts Institute of Technology, at a meeting of the American Chemical Society in Chicago.

The gene is a copy of one present in the bacterium E. coli that orders tRNA

Reprinted by permission of *Science News*, © 1973 by Science Service.

to line up the amino acid tyrosine into protein. Hence the name of the artificial gene is "tyrosine transfer RNA gene." The MIT researchers have not yet completely figured out the start and stop signals on the natural gene, nor completely grafted the signals onto the artificial gene, but they have made ample progress toward this end. Once they get the start and stop signals on the artificial gene, the gene can be introduced into a bacterium via an infectious virus. Once the gene is inside a bacterium, the scientists will be able to see whether it is functional.

Khorana and his colleagues originally began synthesizing a tyronsine tRNA gene with 85 nucleotides. Nucleotides are the chemical building blocks of genes. Then Sidney Altman and John Smith of the University of Cambridge, England, found that there are 41 more nucleotides on the gene. The 126-nucleotide gene was longer than the functional tRNA gene. For some unknown reason, after the long tRNA gene is synthesized, the extra 41 nucleotides break off, creating the functional tRNA gene. So the MIT investigators set out to synthesize the longer gene.

They began building their gene by synthesizing small segments of 10 to 14 nucleotides. Each segment consisted of a complementary portion of two opposing segments of the two-stranded molecule. Thus, each segment acted as a splint to attract and hold together two opposing segments, which could then be tied together by an enzyme called DNA ligase. The scientists designed the synthesis so that these joined segments still had a left-over, single-stranded segment extending beyond the double-stranded segment. This left-over segment was used as a splint to join more segments of the gene. The scientists joined overlapping segments this way until they produced four large portions of the gene. These were joined to produce the entire gene.

As for the stop signal on the gene, Khorana and his colleagues have synthesized 24 nucleotides. They believe these nucleotides comprise a major portion of the stop signal. They are now sequencing the start signal of the gene using the natural gene as a guide.

However important the MIT team's achievements, they are but a beginning in understanding the chemistry of genes. The yeast gene that Khorana and his colleagues synthesized was 77 nucleotides long, the bacterium gene 126 nucleotides. Human genes, in awesome contrast, contain millions of nucleotides each.

Poetic comments on science can also be beautiful.

> imagine being
>   a strand of DNA
> unwinding in the
>   fluid maze
> uncoiling like a breeze
>   while the cloud enzymes
> envelop your backbone
> and the cell wisdom
>   replicates its patterns
> drawn by the needs of bonds
>
> your mosaic
>   a vision of the past
> is locked in chemical messages
> which
>   etch the face of life
> and energy,
> energy, energy
> is lowered and raised
> while
>   the economy of balance
> telescopes
>   the cytoplasmic ballet
> into
>   the forms of the future.
>
>   payson stevens

MOLECULAR RHYTHMS
Payson Stevens

From CRM Books, *Involvement in Biology Today*, copyright © 1972 by Communications Research Machines, Inc. Used with permission.

INTRODUCTION TO BIOLOGY TODAY
Albert Szent-Gyorgyi

> Albert Szent-Gyorgyi is both a Nobel laureate and a scientific rebel. This personal sketch of Szent-Gyorgyi is followed with his unorthodox feelings about heredity.

In my mind I have never been able to accept fully the idea that adaptation and the harmonious building of complex biological systems, involving simultaneous changes in thousands of genes, are the results of molecular accidents.

The feeding of babies, for example, involves very complex reflexes. These reflexes require extremely complex mechanisms, both in the baby and in the mother, which must be tuned to one another. Similar mechanisms are involved in the sexual functions of male and female animals. These mechanisms must be tuned precisely to one another in order to achieve successful copulation. Thousands of genes must be involved in the coding of these mechanisms. The probability that all of these genes should have changed together through random variation is practically zero, even considering that millions or billions of years may have been available for the changes.

I have always been seeking some higher organizing principle that is leading the living system toward improvement and adaptation. I know this is biological heresy. It may be ignorance as well. Yet I think often of my student days, when we biologists knew practically nothing. There was then no quantum theory, no atomic nucleus, and no double helix. We knew only a little about a few amino acids and sugars. All the same, we felt obliged to explain life. If someone ventured to call our knowledge inadequate, we scornfully dismissed him as a "vitalist."

From CRM Books, *Biology Today*, copyright © 1972 by Communications Research Machines, Inc. Used by permission.

Today also we feel compelled to explain everything in terms of our present knowledge. Identical twins are often exactly alike in the smallest details of physical appearance, indicating that the instructions for building this entire structure must have been encoded in the genetic materials that they share. All the same, I have the greatest difficulty in imagining that the extremely complex structure of the central nervous system could be totally described in the genetic codes. Thousands of nerve fibers grow for long distances in order to find the nerve cell with which they can make a meaningful junction. Surely the nucleic acid did not contain a blueprint of this entire network. Rather, it must have contained instructions that gave the nerve fiber the "wisdom" to search for and locate the only nerve cell with which it could make a meaningful connection. Perhaps this guiding principle also is related to the way in which the first living system came together.

I do not think that the extremely complex speech center of the human brain, involving a network formed by thousands of nerve cells and fibers, was created by random mutations that happened to improve the chances of survival of individuals. I must believe that man built a speech center when he had something to say, and he developed the structure of this center to higher complexity as he had more to say. I cannot accept the notion that this capacity arose through random alterations, relying on the survival of the fittest. I believe that some principle must have guided the development toward the kind of speech center that was needed.

Walter B. Cannon, the greatest of American physiologists, often spoke of the "wisdom of the body." I doubt whether he could have given a more scientific definition of this "wisdom." He probably had in mind some guiding principle, driving life toward harmonious function, toward self-improvement.

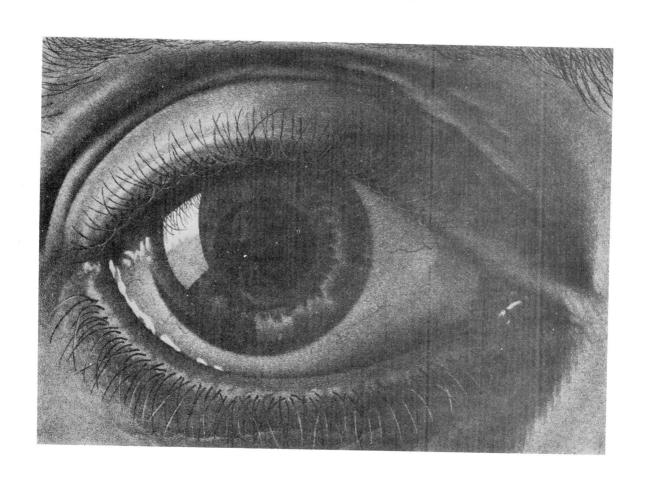

PART VI
STRUCTURE AND FUNCTION

THE LONGEVITY SEEKERS
Albert Rosenfeld

Is aging necessarily a part of life?

We have always accepted old age, with all its attendant miseries, as the inevitable end of human life. Now, however, gerontologists are beginning to question its inevitability. As their researches gradually reveal to them the mechanisms of the aging process, they grow increasingly hopeful that they may one day be able, not merely to hold back the physical ravages of senescence, an outcome most of us would happily settle for, but to abolish altogether "old age" as we have known it.

There is good reason for their optimism. Consider a recent experiment, one of many hundreds in progress in laboratories around the world:

In a surgical technique called parabiosis, an aging rat is hooked up to a young rat so that they share a common blood circulation, like Siamese twins. They are joined, tail-to-shoulder, which leaves them fairly free to move around. For the aging rat, parabiosis acts as a youth transfusion. In studies with more than 500 parabiotic rats, Frederic C. Ludwig of the University of California, Irvine, has found that his older subjects live significantly beyond their expected life-span. Something in the blood of the younger rats enabled the older rats to live long after all their unhooked-up littermates were dead. "When rats are joined parabiotically," says Zdanek Hruza of New York University, "remarkable biochemical changes take place." The older rats' cholesterol levels, for instance, go down "almost miraculously."

Some years ago, at NIH's Gerontology Research Center in Baltimore, Dietrich

Bodenstein used parabiosis to create Siamese-twin cockroaches--again, young joined to old. Young cockroaches can regenerate lost limbs, but old roaches lose this capacity. In parabiotic combination the old roach recovered its regenerative powers. When Dr. Bodenstein lopped off a limb, it was promptly regrown. This renewal was attributed to the transference of juvenile hormone from the young roach to his Siamese twin. Could something similar be happening to the rats? Dr. Hruza says that if a hormone is involved, it's not any of the known hormones. What, then, is the nature of the "youth factor" (or factors) circulating in the blood of a young rat? Could it (or they) be isolated and used to prolong life without the need for parabiosis? And could the same be done for people?

Whatever the specific biochemistry involved, these cases clearly demonstrate that at least some characteristics of old age, formerly considered inevitable, can be held off, as in the case of the rats, or reversed, as in that of the cockroaches.

It has been known since 1917 that reducing body temperature slows the aging process. "In cold-blooded animals," says Bernard L. Strehler of the University of Southern California, "a tenfold increase in longevity has been achieved by lowering body temperature, without adversely affecting body function." Cooling slows things down in warm-blooded animals, too, though not so spectacularly. Recently Robert Meyers of Purdue cooled monkeys a few degrees by direct manipulation of the "thermostat" in the brain and thereby increased their life-span. It has been seriously suggested that people might one day sleep in cooled bedrooms or in slightly chilled waterbeds to achieve longer life.

Dr. Strehler knows of no exception to the rule that animals live longer at lower body temperatures and wonders whether long-lived people might not actually have slightly lower-than-average temperatures. "A very minor reduction in temperature, about three degrees Fahrenheit, could well add as much as thirty years to human life," Strehler says, "assuming, of course, that the human organism operates like its relatives in other species. There is no way to predict the long-term side effects of artificially reduced temperature, but drugs already exist that can reduce temperature by the requisite amount, and others will certainly be discovered.

Techniques such as cooling and parabiosis are comparatively gross methods of combating old age, but they may serve a useful interim purpose while we await further news from the frontiers of molecular biology. Dr. Strehler, a pioneer researcher on these frontiers, typifies the vigorous new breed of gerontologists-- those who study the aging process itself, as distinguished from geriatricians, who study and treat the diseases of old age.

"I really hate death," he says, with a straight face. He speaks with no trace of irony or resignation but rather in the testy, indignant tone one might use to complain about, say, air pollution, or graft at City Hall. "There is no principle in nature," he declares flatly, "which dictates that individual living things, including men, cannot live for indefinitely long times in optimum health."

"Old age is just another disease," says Benjamin Schloss of the University of Buffalo, who is also director of the Foundation for Aging Research in Brooklyn, New York. It is a disease everyone gets, and the individual who survives all the other diseases invariably succumbs to this one. But just because a disease is universal and has always been fatal does not, in Schloss's view, make it inevitably so. Merely to think about aging as a degenerative ailment, rather than as man's eternal and preordained fate, is to put it in the category of a medical problem---something your doctor may some day hope to do something about.

The key to the process of aging may emerge from our new knowledge of the cell. We know that the DNA molecules in the cell's nucleus make up the genes that contain all the basic information for the development and maintenance of life. We have begun to understand how, inside each cell, that information is transmitted via RNA molecules--with the help of enzymes--to carry out the multitudinous, simultaneous activities constantly taking place in the breakdown and buildup of cellular materials, principally proteins. We are, in brief, deciphering the <u>control</u> mechanisms of life. If we really do come to understand these control mechanisms, then the controls may pass to us. If the aging process is governed by these same mechanisms, then we can control aging.

Many scientists now think that aging is genetically programmed. In the original fertilized egg it is "written" in the language of DNA-RNA that we will deteriorate and die. If we take a nontheological view, nature's only interest in our individual health and welfare is to see that we survive long enough to reproduce ourselves and to raise a new generation that will do likewise. Once the propagation of the species has been assured, nature is ready to have us make room for newcomers. DNA, the immortal molecule, renews itself through the continuing generation.

An extreme example of the aging process is the Pacific salmon. As it comes in from the ocean, laden with eggs, headed for its freshwater spawning grounds, it is an impressively beautiful fish--a shiny, orange-red creature powerful enough to churn far upstream past rapids and waterfalls. Once it has safely deposited its eggs, however, the salmon deteriorates with appalling swiftness, speeding from youth to senility in scarcely two weeks.

Humans undergo a similar degenerative process after we have passed our reproductive and child-rearing years. But nature gives us (or tortures us with) a longer time to decline. The skin begins to wrinkle and spot. The hair turns gray and thins out. The teeth decay; the eyes grow dim and the ears dull. The muscles get weak and flabby. We all recognize the stooped and shuffling gait that accompanies old age. Meanwhile, internal deterioration also sets in, probably much earlier in life than is apparent from the outward signs. The lungs can't take in as much air as they used to. The heart pumps less blood. The arteries become coated and hardened. The liver and kidneys perform their functions with diminishing efficiency.

The intellect, too, deteriorates. Concentration and learning ability are reduced. After the age of about thirty-five our brain cells die off at a rate of roughly 100,000 per day from an initial total of ten billion or so. Fortunately, the diminution of mental powers does not usually come about until quite advanced ages and is often counterbalanced by wisdom acquired through experience. (Many experts believe that we have greater brainpower, in any case, than we ever make use of.)

Brain cells, once lost, are never replaced. Nor are some of the other cells --those of the heart muscle, for example. But most of the body's cells do replace themselves constantly, simply by continuing to divide whenever replacement is necessary. Even among the dividing cells, though, there is some loss. Nearly all cells lose some of their capacity to retain vital fluids and eliminate waste matters, so they may be come clogged and sluggish. Some of them turn into connective tissue, which gets tough and fibrous, like vulcanized rubber. Hence we get drier and stringier as we age. And there is a very substantial loss of tissue--including bone tissue, especially if the individual falls victim to osteoporosis, a fairly common disease of old age. As we grow old, we actually shrink, becoming appreciably shorter and lighter. Moreover, with advancing age, the entire organism becomes less able to deal with stressful situations--and recovers from them more slowly. It loses much of its reserve capacity and falls more easily prey to degenerative diseases--cancer, heart disease, atherosclerosis, hypertension, diabetes, arthritis, senile dementia.

Old people are <u>expected</u> to get sick. A doctor's standard comment to a patient, anywhere from thirty-five to ninety-five, is "You're in pretty good shape <u>for your age</u>." Over the years of this century, with the decrease in infant mortality and the conquest of many infectious diseases, more and more people live to be old. The average life expectancy has been considerably increased, but people who survive to old age don't live any longer than they used to--and often are kept going only through constant medication. Some 10 per cent of the American population is now over sixty-five, and, of these, 86 per cent have one or more of the degenerative diseases.

But no one ever emphasized the reverse side of these statistics: 14 per cent <u>don't</u> have these diseases. Many old people seem to escape the major ravages of aging until they are very old indeed. Why is it that one man is bald and toothless twenty years sooner than his next-door neighbor? What keeps a Picasso or a Casals vigorously creative into his eighties and nineties? Or a statesman like Churchill or Adenauer? Or writer-philosophers like Bertrand Russell and George Bernard Shaw? Right now there are 13,000 people in the United States--more of them women than men--who are over 100 years old. Are they just lucky, or is long life written into their genes?

Leonard Hayflick of Stanford University compares the human organism to a vehicle launched by NASA to fly by Mars and relay information back to Earth. The engineers worry about the soundness of all the working systems--sensors, radio transmitters, rocket engines--all the way to Mars. Once the mission is completed, however, they have no further interest in the vehicle, though it does continue to fly through the universe. How long does it last? Which parts fail first? The experimenters don't care. That's the way nature and evolution seems to "feel" about us, once our propagative function is achieved.

Unlike the Mars spacecraft, however, <u>we</u> care what happens to us--and we have independent brains to think with. We have already used "unnatural" means to combat other ailments (God didn't furnish us with scalpels and antibiotics), and now we may discover how to combat the ultimate disease--old age.

Dr. Hayflick's own research may help speed us toward that discovery. It was once believed that cells grown in laboratory tissue cultures were essentially immortal. That is, as long as they were given adequate nutrients and a congenial environment, they would go on dividing forever. But Dr. Hayflick proved that this applied only to abnormal, usually malignant, cell lines. Normal cells have a finite limit. A given normal cell strain taken from an embryo might divide fifty times, for example, and then stop. The older the individual from which the cells were taken, the fewer times they divide.

More and more, it appears that there is a maximum life-span programmed into human genes. If an embryonic cell strain divides, say, twenty times and then is put into the deep freeze for months or even years, it "remembers" where it left off. After thawing, it divides another thirty times--but no more.

Cells taken from the victims of the rare disease called progeria divide only a few times. A child afflicted with progeria runs through his entire life cycle at a strangely accelerated pace. He may begin to show signs of aging by the time he is a year or two old. He may be an old man by seven or eight, shuffling along bald-headed, rheumy-eyed, and heart-troubled, and then die at the age of eleven or twelve.

No one is yet positive that progeria represents a true aging process or that it is genetic in nature, though it does appear to be. But a similar disease, Werner's syndrome, is which the aging acceleration doesn't begin until the teens, is definitely known to be a single-gene disease. If the genetic program can be speeded up by mistake, can it be slowed down on purpose? William Reichel of the Franklin Square Hospital in Baltimore, who is making a study of these ailments, does not consider that a foolish question--though he is not yet close to having the answer.

Neither in a victim of progeria, where the clock of aging seems to run wild, nor in a normal person, are we sure whether the organism as a whole deteriorates, therby causing changes in the individual cells, or whether senility and death are merely the visible reflection of what takes place in billions of cells as they slowly deteriorate. Both interpretations may be correct. But what happens in the cell is of critical importance.

Some scientists speculate that as cells age, there is a "blurring" or damaging of the genetic material--that they begin to divide imperfectly or not at all. Even in cells that do not divide, there is a lowered capacity to perform vital functions. This almost certainly means a loss of, or change in, the genetic information, a loss or change that may occur in the DNA itself, perhaps through mutations caused by the chance impact of cosmic rays. Or it may be a loss in the information-carrying, or information-copying, capacity of molecules that transmit the genetic messages or that manufacture the protein. Leslie Orgel of the Salk Institute believes that almost any error, if it occurred in a critical enzyme, could result in a chain reaction of errors--an "error catastrophe" that would interfere with, or altogether stop, protein manufacture. John Maynard Smith of Sussex University in England has suggested that cells might contain certain long-lived, one-time-only proteins--which, when they run down or wear out, are irreplaceable.

Some researchers believe that trouble is due to "cross-linkage" of large molecules that, as the years go by, accidentally become tangled with one another. Others are convinced that "auto-immunity" is a major cause of aging--that the body's natural defenses against disease begin to attack normal cells--either because the information is blurring or because the other cells are changing in ways that make them look "foreign."

Another suspected aging agent is the presence in the cell of "free radicals" --which might be characterized as pieces of molecules eagerly seeking molecules to latch onto. "A free radical," says England's Alex Comfort, perhaps the world's most respected gerontologist, "has been likened to a convention delegate away from his wife: it's a highly reactive chemical agent that will combine with anything that's around." These combinations can have damaging results. They might cause cross-linkage. They could disrupt the information content of important molecules. They could accelerate the formation of hard-to-dispose-of cellular garbage, especially a dark pigment called lipofuscin. Large quantities of free radicals are produced constantly by the cell's own oxidation reactions.

The experiments in cooling and parabiosis, already described, prove fairly conclusively that there is nothing immutable about the rate of aging. This conclusion is reinforced by other recent laboratory research in gerontology:

X When young skin cells are transplanted to older animals, they grow old with the animal; but young skin cells transplanted to a young animal and constantly re-transplanted to other young animals live six times as long as they would have had they been left in the original animal. What do these skin cells get from their hosts that extends their life-span so impressively?

X The rotifer is an aquatic animal tinier than the period at the end of this sentence. Its normal life-span is eighteen days. Charles H. Barrows, Jr., of NIH's Gerontology Research Center in Baltimore has found that reducing the water temperature by $10°$ C. will almost double the rotifer's life-span. If, at the same time, the amount of food is cut in half, the life-span triples. But different periods of life are affected: food restriction lengthens only the early part, while lowered temperature lengthens only the latter part.

X Experiments conducted years ago at Cornell by Clive McKay also showed that restricting caloric intake prolongs life (in this case, the life of rats)--but again, only the early period of life. A similar stretching out of the earlier life --of mice--was achieved by Denham Harman of the University of Nebraska, who was using antioxidants to combat free-radical damage.

X Roy L. Walford of UCLA has tested the auto-immunity theory by using on mice drugs that suppress the immune system and, by so doing, has apparently been able to retard the aging process.

X Female hormones seem to protect the heart muscle in rats from deteriorating (human females, it may be noted, have a much lower risk of heart attack than do men) and can definitely rejuvenate certain tissues in the reproductive tract of aging women.

In the course of hundreds of gerontological experiments, measurements have been made of the specific changes in cellular function that take place with age. Many enzymes in animal tissues grow less active with age: others, such as LDH (lactic dehydrogenase), increase their activity. Mature DNA has a more tightly folded structure than younger DNA. The RNA content of chromatin increases with age. The liver, lung, and prostate cells of old rats can't synthesize RNA as well as they used to. There are fewer kinds of RNA present as cells age--but some entirely new types appear. Protein manufacturing varies considerably, as does the ratio of DNA to RNA. Antibody production is lower, but the auto-immune response is higher. There are myriads of such changes, few of them well understood, but just about all of them governed by genes.

Can anything yet be done about reversing or interfering with genetic information? Very little at this time, but it is clear that someday we will be able to do a great deal more. If the critical changes are in the RNA molecules or enzymes of the cytoplasm, rather than in the nuclear DNA itself, the task will be more manageable, because the cytoplasm is more accessible than the DNA. At the University of Buffalo, J. F. Danielli and A. Mugleton took a species of amoeba that is essentially immortal and, by transferring cytoplasmic material from a mortal species, conferred upon it the gift of old age and death. There is no reason why refinements of this technique could not accomplish the opposite result.

Tinkering with the DNA presents greater difficulties, but they are not insurmountable. One set of experiments has provided evidence that the decline of RNA manufacture (a process directed by DNA) in the liver of an aging rat could be reversed by the addition of fresh DNA from a young rat. Even more interesting are a couple of recent attempts to remedy human genetic defects:

X In a case described as the first attempt at "genetic engineering," two young German girls suffering from a genetic disease were started on a radical course of treatment. Because their ailment is characterized by low blood levels of the enzyme arginase, their doctors in Cologne gave them injections of a virus that is known to produce very high blood levels of arginase. No one yet knows if the treatment will work.

X In a 1971 experiment at the National Institutes of Health, scientists took skin cells from a patient with galactosemia, a sometimes fatal genetic condition that prevents the body from metabolizing milk. They used a virus called lambda phage, which possesses the missing milk-processing enzyme, and were able to transfer this genetic information via the virus to the skin cells so they could begin to do the job themselves. In a word, they succeeded in artificially imparting the information necessary to correct the inborn genetic error! This is exactly the kind of subtle genetic tampering that may have to be done to correct aging changes.

Fortunately, given the many thousands of genetic changes going on simultaneously at every step of aging, we now know that some substances serve as regulator substances that control the clock of aging. Some gerontologists believe that such clock-controlling "pacemaker" chemicals do exist and that they may be depleted or damaged as time goes by--or perhaps programmed to disappear at given times. The Pacific salmon's instant deterioration may be related to such a disappearance. In an insect the mere release of a hormone at the appropriate time can trigger an incredibly well-orchestrated sequence of transformations, so radical that the metamorphosed result appears to be an entirely different creature. The hormone orders thousands of events in millions of cells, many of which set off their own built-in self-destruct mechanisms. A scientist who knows how to administer such a hormone doesn't have to understand all the details, any more than we need to understand electrons and tungsten filaments in order to turn on a light.

Searching for such a hormonal solution is W. Donner Denckla of the Roche Institute of Molecular Biology in Nutley, New Jersey. He believes that the clock of aging resides in the brain--specifically in those cells that govern the release of certain hormones. Dr. Denckla is looking for a still-undiscovered "death hormone" or family of hormones whose nature he has theoretically calculated. Released at critical times of life, the death hormones may progressively inhibit cells from utilizing other hormones, probably thyroid hormone. The reason for this suspicion is that hypothyroidism--deficient thyroid activity--closely mimics aging. Its symptoms are reversible through thyroid administration, but only if the patient is young enough. After a certain age people do not seem able to take up thyroid hormone, even if it is present in the bloodstream.

Suppose a scientist already had in hand what he believed to be the pacemaker chemical or chemicals that control the clock of aging--or perhaps, as an interim measure, a "cocktail" containing many of the chemicals known to be depleted with age. He would naturally want to test it on people. Wouldn't it still take many generations, and the use of vast populations, to prove that it was really capable of prolonging life? This has always been one of the discouraging inhibitors to gerontological research in humans: the researcher can't see the results of his work in his own lifetime.

However, this may be changing as medical technology advances. There already exist automated blood analyzers that can, with a single drop of blood, quantify a dozen or more different blood characteristics within a minute or two. Large-scale versions of these are now being developed. If successful, they will be able to analyze hundreds, even thousands, of different substances at once. Dr. Schloss has attained a patent on such a device, and another, based on ultraviolet spectrography, is being developed by Dr. F. J. G. Van den Bosch at the Downstate Medical Center in Brooklyn, New York.

If these devices work as hoped, a single blood sample could tell us a great deal about the functions of the liver, kidney, and other organs in terms of the chemicals they put into the bloodstream. If we can list the significant biochemical changes that take place during aging, then we can measure them all at once in a single person.

By doing this on many persons of different ages over a period of time, we can begin to establish a "biochemical profile" of a person whose average age is, say, thirty. It would also be possible to measure quickly his _rate_ of aging over short periods of time. Then one can readily tell if a pacemaker chemical or gene tampering is achieving the goal of prolonging youth.

"This approach," says Dr. Comfort, "reduces the problem of how to retard aging in man to the size of an ordinary medical experiment, using some 500 volunteers over three to five years, like the assessment of low-cholesterol diets in heart disease."

There are some things gerontologists do _not_ especially want to do:

X They have little interest in stretching out the years of senility, to keep people alive in a state of helpless decrepitude—though they would certainly like to relieve the aches and anxieties of this period and, if possible, undo some of the damage.

X They do not want merely to prolong the years of childhood, as has been done in rats and rotifers through restrictive feeding—though the ability to achieve this might have some interesting possibilities.

X They do not seek to "prolong" life by inducing periods of hibernation during which a person would continue to exist in a state of suspended animation—like Dr. Hayflick's deep-frozen cells—to be awakened for intermittent periods of active life. This might be a fascinating experience, even a useful one—especially for long space voyages—but the total years of wide-awake life would remain three score and ten.

What the gerontologists _do_ want to do is to extend the number of continuous years spent at a productive and enjoyable level of health and intellectual power. Presumably no one would be required to go on living, but those who wanted to would have the option. Given a real choice of prolonged _good_ years, how many would turn it down?

SHUFFLE BRAIN
Paul Pietsch

Paul Pietsch brilliantly describes on the most abstract concepts of contemporary biology--the holographic theory of mind.

Punky was a salamander. Or at least he had the body of a salamander. But his cranium housed the brains of a frog. I'd spent an entire season at the fringe of his clear-water world, asking who he was, with the neural juice of a totally different animal racing around inside, turning him on, tuning him in to his environment at a wave band beyond a normal salamander's spectrum. The answers, borne by his actions, flattened my scientific detachment, I confess.

Punky was only one in a long and varied series of brain transplants, experimental tests of the holographic theory, a theory about the language of the brain, a scientific treatment of nothing less than memory itself--the watering hole on the great subjective plain where thoughts and dreams, hopes and fears, pride and guilt, love and hate must drink to live, or else dry up, to vanish, like bone dust.

Years before, in Philadelphia, when I was first learning how to do operations like those on Punky, I was an instructor in a gross-anatomy dissecting lab. Class met in the afternoon. Insecure in my grip on what was then a newly acquired subject, I went in early each morning to do a dissection of my own. With class in session, the place roiled with the hurly-burly of people, alive and busy. But in the morning, when I arrived, it was silent, a room of death in the most complete sense of the word. Ugly gray light glared in through frosted windows and, without color, illuminated the rows of rag-swaddled, tarp-wrapped cadavers. It wasn't

---

Copyright 1972 by *Harper's* Magazine. Reprinted from the May 1972 issue by special permission.

frightening; it was lonely, the loneliest place I'd ever seen. Its tables were the biers of the world's unwanted, unremembered, unclaimed--as people. And they'd been forgotten long before their corpses were hoisted up and flopped naked on the diener's soapstone prep table. Nameless now, serial-numbered metal-ring tag tied around big toe, dirt still under cracked nail or maybe half-peeled-away red or pink nail polish. Valuable, in death, as things. Valueless before, as people. They were the unloved dead. For to be loved is to be remembered. They were the unhated dead, for the same abstract reasons. The unremembered dead, the truly dead. For memory is our claim to identity, and when it stops, we are no more.

At the end, when we were finished, my department held funeral services for the bodies. I went. But I went with a generalized grief that I carried back whole because my memory found no place to assign any part of it.

Still, in time, I did forget the details. But Punky revived my memories of those mornings back in Philadelphia. That's probably why I gave him a name. For the Existenz of Punky and his pals didn't stop with salamanders and frogs. It included my own species.

CUTTING AWAY THE MIND

I will be talking here about the neural hologram, but I really should speak of brain information--a hologic principle, not only memory of past experiences. For the theory seeks to explain all the brain's stored programs, whether learned or wired in during embryonic life. It covers the mental yardgoods we unwrap to tailor "go: no-go" in reflexes. It supplies the cash for complex, reasoned associations. It works when the brain issues instructions to tune the A-string on a viola, or to make the baby cry because the milk is sour.

But holographic theory deals with the mode of neural messages, not specific molecules, mechanisms, or cells, as such. Like a multiplication table or counting system, it commits grand polygamy with place and time and circumstance. It treats the how rather than the who--like gravity acting on the apple, instead of the meat, the freckles, or the worm.

The holographic theory had its crude origins in the 1920s when psychologist Karl Lashley began a lifelong search through the brain for the vaults containing memory. By then, students of behavior had been readied for angry debate by a paradox that had begun to emerge on the surgical tables of the nineteenth century. Clearly, the mental world had its biological base in the brain. Yet war, disease, and the stroke of the scalpel had robbed human brains of substance without necessarily expunging the mind. Lashley carried the problem to the laboratory and pursued it with precision tools, mazes, rats, controls, statistics.

Lashley also brought along the knife. With it, he found he could dull memory in proportion to the amount of cerebrum he cut out. But if he left a rat with any cerebrum at all, the animal could still remember. Not only did he fail to amputate memory, but one area of the cortex would serve it as well as another. He came to two controversial conclusions: intensity of recall depends on the mass of brain, but memory must be divvied up equally. "Mass action" and "equipotentiality" became his theme.

"Equibull!" a neuroanatomist friend of mine once declared. For the knives and battery poles of others had struck and dug into what seemed to be the specific loci of sight, scent, sound. Moreover, no clear and obvious physical precedent existed for equipotentiality. "I'm a scientist" my friend used to say, "not a goddamn Ouija board operator!"

But in 1948 physicist Dennis Gabor, trying to improve the electron microscope, accidentally stumbled over the optical hologram, a discovery that earned him the Nobel Prize in 1971. Lensless, 3-D photography was born. Within twenty years, the same principles had been extended to the brain.

Holograms take getting used to--like the idea that light can be both waves and particles, or that a curve gets you more quickly from star A to star B than Euclid's straight line. It's like getting accustomed to the notion that energy and mass are different ways of saying the same thing, or that time might shrink and expand. For holograms package information in a form disguised from our common sense, invisible behind the nominalistic curtains of our culture. But with patience, and a little open-mindedness, the intuition soon begins to drink up the principles--like relativity after Einstein or the shape of the earth after Columbus.

Familiar modes of information, even as complicated codes, reduce to bit parts, held, stored, according to the <u>summum bonum</u> of home economics and gross anatomy: "A place for everything, and everything in its place!" Not so a hologram (<u>holo</u> means whole). In it, the entire <u>shtick</u> of information, tamped down into a <u>minuscule</u> transcendental code, repeats itself, whole, throughout whatever the system happens to be. Trim a hologram down to a tiny chip and the message still survives, whole, waiting only to be decoded. One piece will work as well as another. But the fewer the parts used in decoding, the less intense the regenerated image. In other words, holograms work in precisely the same way that the memories in Lashley's rats did--mass action and equipotentiality.

Gabor's discovery was for years of scientific curiosity, unknown outside a small circle of physicists. It remained so until the advent of laser technology. Holograms in physical media depend on coherent, orderly waves. To do anything other than just look at holograms, the waves must be fairly powerful. Laser beams not only have this property but they can be made very coherent.

Holography itself has bloomed into a new technology. There are even such people as holographers nowadays. They construct physical holograms for a living, and they are paid well to do so because the hologram may be the method of sending and storing information in the future.

To construct a physical hologram, a holographer uses two sets of waves. He shines oue set through an object. He angles the other to miss the object but to collide with the waves that have passed through. He then collects the results of the collision on film or a cathode ray tube. His record, the hologram, represents the reaction between the distorted and undistorted waves.

In appearance, physical holograms resemble Platonic <u>ideas</u> of a shivering tiger or zebra or the signature of an artist suffering the shakes of a bad whiskey hangover. But the holographer can regenerate an image of the original object by retracing his construction procedures.

A hologram captures not a thingy thing. It captures rules--a harmonic syllogism, a hololo gic. And it is the stored record of Hegelian skid marks produced when points and counterpoints bang into each other, physical or numerical, concrete or abstract. Mathematics in reverse. Indeed, they take getting used to. But the glory of holograms glows through during decoding back to the original image, when they not only behave like Lashley's rats but reveal feature upon feature of human brain function.

Holographers can construct, say, acoustical holograms and call back the original, not with sound, but with light or waves in some other form. Thus, built into holographic grammar is the automatic mechanism to shift gears, instantly, from one modality to another--how, for example, you can listen to someone and <u>write</u> what you <u>hear</u> him say as fast as you can work the muscles in your hand.

Such rapid, whole-scene shifts, involving forests of data, would be out of the question with the conventional message that must be translated bit by bit. In a hologram, it's all part and parcel of the principle. And the same thing shows up again in adding and modifying holograms. Holographers can construct multicolored, composite holograms, in steps, by adjusting wave-length, thus mimicking how we might anneal present and past into a totality. Or they can decode several holograms of the same thing into a multicolored original. In the process they can even change colors. When the brain does these things on its munificent scale, we talk in terms of abstract reasoning or imagination. And in this capacity the human brain outshines the largest digital computers. For computers digest bits. But the brain's motifs are informational wholes that can meld and blend without the go-between of a finger-counting bureaucrat.

The flexible rules of holography even allow, automatically, for a subconscious, a bad word in my own particular profession. But consider an optical hologram. In decoding, it's possible to select a wavelength invisible to the naked eye, yet of sufficient energy to burn a monogram permanently onto the retina of an unwary onlooker. As with the subconscious, you don't have to see its wounds to ache from them.

Holographic theory would also explain the chemical transfer of memory—how information from the brain of one worm, rat, mouse, or hamster might be extracted into a test tube and injected into another animal, there to mediate recall in the absence of the recipient's previous experience. Such reports from a dozen laboratories over the past few years have excited the press and reading public. But in conventional scientific circles, I've heard them called such things as "oozings from the stressed seams of cracked pots." Yet a hologram can write itself into anything, including a molecule. At the very same time, the theory in no way at all restricts the brain's programs to molecules, as such. There's no rule against using, say, molecules, voltages on cells, or groups of neurons to carry the information. The program might even be carried at many different levels simultaneously.

Just who deserves credit as the first to apply holographic principles to the brain I'm going to allow historians of science to fight out. Lashley, of course, saw them at work in his rats and had both the genius and the courage to describe what nature showed. Certain of Pavlov's conclusions look hological. Gabor's powerful mind must have snared the notion the moment he tripped on the optical effect. Years later, in fact, he published a mathematical scheme of reminiscing. Philip Westlake, a brilliant UCLA cyberneticist, has shown that equations of physical holograms match what the brain does with information. Karl Pribram and an army of colleagues at Stanford's medical school have invested a decade and a thousand monkeys, using the theory to work out details of how living brains remember.

Predictably, holographic talk provokes hot controversy. I recall not long ago delivering a lecture on the subject, when out of the audience jumped a neuropharmacologist, trembling with rage, demanding to know: "How can you account for something like Broca's area?" He was referring to a part of the cerebrum known for 100 years to be vulnerable to stroke accompanied by the loss of speech. I cleared my throat to answer. But before I had the chance, a young psychophysicist sprawled in a front-row seat, whipped his shoulder-length mane around and fired back, "You can't draw beer out of a barrel without a bung!"

It was a perceptive reply. For in holographic theory, functional centers such as Broca's area represent processing stations rather than storage depots. Rage, fear, hunger centers, the visual cortex at the back of the brain, or auditory areas at the sides--these would act not to house specialized information but to pump it in or to call out programs in the form, say, of snarl, smile, utterance, equation, kiss, or thought. And sharp lines of distinction between innate and acquired information fade as far as storage itself is concerned.

Still, the theory does not completely rule out uneven distribution of memory, particularly in the complex brains of higher animals. Indeed, it is not hard to make a case for different storage within the two hemispheres of the human cerebrum. Michael Gazzaniga recently published an intriguing book on what has been known for almost twenty years as "split-brain" research. Begun in the early 1950s by Meyers and Sperry at Cal Tech, the technique involves cutting the corpus callosum, a broad thick strap of nerve fibers between the hemispheres. Success in the lab with cats and monkeys prompted neurosurgeons to split the corpus callosum in the human brain. They did so to alleviate violent, prolonged, drug-resistant grand mal epileptic seizures, and they had remarkable success, medically. But the patients emerged from surgery with two permanently disconnected personalities. With more such operations, the left cerebral hemisphere emerged as the dominant, verbal, arithmetic side, which the right brain held recollections of form and texture. The tendencies appear to hold whether patients were left- or right-handed. Early in 1971, music was found among the repertoire of the right hemisphere. Yet the outcome of split-brain surgery has never been absolute, nor the individual patient's subsequent behavior totally predictable. Both hemispheres can generate music in some people, and the right may have a vocabulary. In addition, a totally illiterate right hemisphere can learn to read and write in less than six months--as though it had a tremendous head start. On top of this, Gazzaniga's observations convince him that the consignment of memories to one side of the brain emerges with maturity. Children seem to employ both hemispheres. Thus it would seem that the brain can reshape its contents and make decisions about what will go where. But it is also quite possible that split-brain research identifies not unequal storage but unequal access. Like the reflected image of a written message, meaning would stay the same but translation would entail different steps. The cerebral hemispheres, after all, do mirror rather than carbon-copy each other.

At any rate, the brains of human beings and our close relatives seem to be many brains, orchestrated by virtue of connections like the corpus callosum. Moreover, our multisystem cranial contents seem to be in flux, physiologically. Different lights can flash off and on, moment to moment. Some of the switches lie under our direct control; others are no more within our deliberate, intellectual reach than the impulses driving a hungry shark or an amorous jackrabbit.

Holographic theory does not deny conclusions of split-brain research. But it insists that, whatever the system used for storage, the information shall be layered in whole and repeated throughout. It denies that memory depends on minced-up and isolated bits filed in specific pigeonholes. Just what happens to be going on inside a brain when it's loading up with a particular hologram may determine which areas may and may not act as targets--or how vivid the reconstructed scene becomes during some later translation into conscious form.

THEORIES AND EXPERIMENTS

It's one thing, though, to use a theory to draw complex sets of data or weird collections of observations into a larger body of knowledge. It's quite another to subject a theory to logically valid, epistemologically sound laboratory experiments. Personally, I think it's legitimate to employ a theory without really bothering with formal tests and even to nurture belief in its truth based on its usefulness. For theories supply powerful intellectual tools. Those that don't work very well become ornaments if they're beautiful, junk otherwise.

However, the theory that holographic principles could account for neural-information storage was testable. Before getting into those tests, we need to talk some about theory and experiments. For they belong to very different realms.

Theories are perfectible and can be made ubiquitous within confines set down by their inventor. When they try to say something about physical things, they reach for the harmony and simplicity in nature, for a side of it the inveterate experimentalist believes he can comprehend by observation. To some biologists, for example, the Cell is a fiction. Only cells exist. At best, the theorist regards experience as a start on the road to truth, as Einstein did. At worst, the theorist might tell you that God contrived experience to pollute man's view of the truth. Whatever the experimenter concludes, the theorist seeks an ever-larger synthesis. For the particulate, nominalistic character of an experiment means that it cannot extend far enough and spread out wide enough to cover the expanse of a theory. Thus explanation demands theory.

Even so, because experiments take place in experience, they keep the experimenter alive to a side of nature that theory misses--its variety, its individuals. Theory would turn nature into a peneplain, smooth, unenriched, simpler as the theory reaches higher and higher abstraction until even a speck of dust would become something to cherish. Experience returns hair, lips, smiles, surprises. Experience is where doves coo, horses snort, and robins lay little blue eggs. We spend most of our time, mentally, where experiments go on. And if there is some harmonious thread weaving through the universe, we still have a right to want to connect the abstract world and the world we call real.

How is this done? By poking around in theoretical constructs for testable predictions: if such and such a theory is "true," then such and such an outcome will happen. This is how experiments come in; they are ways of setting a trap for the predictable elements of a theory--the parts of it that make the rules credible to the human mind.

My purpose in working with Punky and his pals was to make or break my faith in the holographic theory of neural storage. And I was a skeptic, at the outset.

When I began this work the only prima facie experimental evidence to link the general theory involving holographic principles to brains had come from ablation studies--subtracting from brain substance. Subtraction is an incomplete test. To see the incompleteness is to see how the salamanders relate to the theory. Thus, let's spend a little time doing a few imaginary experiments.

Imagine several hundred Xerox copies of this unholographic page, but reproduced on transparent plastic sheets. Now stack the sheets so that each letter, word, and line forms a perfect overlay with its replicates below. Now subtract a sheet--two, three, or any number, for that matter--only keeping the stack straight. What happens? Loss or unevenness in density, perhaps. But as long as we keep the <u>equivalence</u> of one page, we preserve the message. The reasons are obvious. First, we're working with a system containing a redundant message. Secondly, when we eliminated some parts, we merely allowed what was beneath to shine through. But we certainly don't have a holographic system. This is how I viewed the results of ablation studies.

Let's try another series of experiments with the transparencies. Let's throw the pile up in the air, arrange some of the sheets in a new order, cut some of the sheets into pieces and reglue the pieces randomly--reshuffle, in other words. Now we would distort the message and know it very quickly. Why? <u>Meaning</u> in a conventional message (or pattern) depends on relationships _among_ parts and subparts--sets and subsets. When we scrambled relationships, when we messed up the system's anatomy, we wrenched the carriers of meaning. We might also have done this by adding a transparency with a different message. But when we merely took away parts from our redundant system, we created empty sets and voided rather than distorted relationships.

But suppose the linotype operator had set a hologram? Then our reshuffling experiments would have produced far different results. We would not have introduced changes in the meaning of the message. For in a hologram meaning lies _within_--not among--any sets we might produce by simple physical means. And in _reshuffling_ we would be shifting whole messages around, exchanging their positions without really getting at components. Trying to dissect out a hologram's subunits is like trying to slice a point, or stretch that infinitesimally small domain by an amount no larger than itself. No, a knife won't reach inside the heart of a hologram. Of course, in practice we might trim a system to such small proportions that the image upon decoding would be too dim to register. Or in a physical experiment we could destroy or distort the medium and make it technically impossible to decode. That's why we opted for imagination--to bypass engineering details.

But look at the implications of our imaginary experiments. Look at the predictions. If we really want to test holography against redundancy, we ought to shuffle the brain. If it houses conventional messages, we would find out very quickly. But if programs exist in the brain according to holographic principles, scramble though we may, we won't distort their meanings. And that is where salamanders come in.

BRAIN TRANSPLANT

A peaceful, quiet world, the salamander's--unless you happen to be a dainty little daphnia or a cockeyed mosquito larva whiplashing to the surface for a gulp of air. Or even worse, the crimson thread of a tubifex worm. For it is the destiny of the salamander to detect, pursue, and devour all moving morsels of meat small enough to fit inside his mouth. He eats only what moves. And he adjusts his attack to fit the motion of his fated quarry. When he sees the tubifex worm, or picks it up on sonar with his lateral line organs, he lets you know with a turn of his head. Position fixed, half-swimming, half-walking, he glides slowly, deliberately, along the bottom of his dish, carefully not to create turbulence that, in the wilds, would send the worm burrowing deep into the safety of the mud. Reaching his victim, he coasts around it, moving his head back and forth, up and down, to catch swelling and shrinking shadows and vibrations and permit his tiny brain to compute the tensor calculus of the worm's ever-changing size.

The size of a four-year-old's little finger, salamanders sustain injury and recuperate like few other creatures on earth. Consider, for example, what I call the Rip Van Winkle paradigm. Remove a salamander's brain. The behaviorally inert body continues to live, indefinitely. Transplant the brain to the animal's broad, jelly-filled tail fin for storage. After a month or two, slide the brain out of the fin and return it to the empty cranium. In a couple of weeks, after the replant takes, the animal behaves as if the operation had never occurred. He's awake again, a free-living, prowling organism, like his normal brothers and sisters.

That same tail fin will accommodate hunks of brain pooled ad hoc from several different salamanders. The pieces quickly send out thousands of microscopic nerve fibers that weave a confluent network. Does such a mass of brain tissue work?

Communicate impulses? Splice a length of spinal cord on each end of the mass as a conduit to the skin. Then, on one side, graft an eye, pressing the cut optic nerve against the piece of spinal cord. On the other side transplant a leg, making sure that it touches the conduit. Wait a couple of weeks to allow the optic nerve to invade the spinal cord on the one side and the cord on the other to sprout fibers into the leg to reinnervate its muscles. Now aim a spotlight at the tail and focus on the grafted eye. If you can hit the light switch at the correct tempo, you can make the transplanted leg stomp a tarantella.

Yet if my experiments were to be a fair test of the holographic theory, I'd have to insure two things. First, the experimental salamander would have to be capable of sensing a tubifex worm. Secondly, he'd have to be able to command his body and jaw muscles into action. I was sure this could be done with salamanders by preserving the medulla, the transitional region between spinal cord and the rest of the brain. In the medulla lie input stations for touch from the head, the salamander's efficient sonar system, and the sense of balance from a carpenter's level-like internal ear. Also, impulses that bring jaw muscles snapping to life are issued directly from the medulla. It does for head muscles what the spinal cord does for, say, the biceps or muscles in the thigh. And in salamanders the medulla serves as a relay station for information to and from spinal cord and brain. Higher animals have such stations too. But evolution added long tracts that function like neural expressways.

There are actually five main parts of the brain common to all vertebrates, including man. The cerebral hemispheres that predominate within our own heads are small lobes on the tip end of a salamander's brain. But during embryonic life our own cerebral hemispheres pass through a salamander stage.

The next region back, known as the diencephalon, is where the optic nerves enter the brain. Distorting this region would and did create blindness in certain experiments. A so-called mesencephalon or mid-brain connects diencephalon to medulla. These were the parts I would shuffle.

Amputating brain in front of the medulla turned off the salamander's conscious behavior and, of course, feeding along with it. But, if I stayed out in front of the medulla, I'd be leaving sufficient input and output intact for whatever programs surgery might deliver up.

This is not surgery in the nurse-mask-sutures-and-blood sense. It goes on under a stereoscopic microscope. Very little bleeding. No stitches. Just press the sticky, cut tissues together and permit armies of mobilized cells to swarm over and obscure the injured boundary line. There is only room in the field of operation for a single pair of human hands. The animals sleep peacefully in anesthetic dissolved in the water. Trussed lightly against cream-colored marble clay, magnified, they look like the prehistoric giants of their ancestry. A strong heart thrusts battalions of red blood corpuscles through a vascular maze of transparent tissues. No bones to saw. Under fluid your instruments coax like a sable-hair brush.

In more than 700 operations, I rotated, reversed, added, subtracted, and scrambled brain parts. I shuffled, I reshuffled, I sliced, lengthened, deviated, shortened, apposed, transposed, juxtaposed, and flipped. I spliced front to back with lengths of spinal cord, of medulla, with other pieces of brain turned inside out. But nothing short of dispatching the brain to the slop bucket--nothing explunged feeding!

Some operations, created permanent blindness, forcing animals to rely on their sonar systems to tell them what was going on outside. But the optic nerves of salamanders can regenerate. Still, for normal vision to return, regenerating optic nerves need a suitable target, as Roger Sperry showed many years ago. I was able to arrange for this, surgically. And when I did, eyesight recovered completely in about two weeks--even when the brains came from a totally different species of salamander and contained extra parts. As far as feeding was concerned, nature continued to smile on holography. Not one single thing about the behavior of this group of animals suggested the drastic surgery they had undergone.

The experiments had subjected the holographic theory to a severe test. As the theory predicted, scrambling the brain's anatomy did not scramble its programs. Meaning was contained within the parts, not spread out among their relationships. If I wanted to change behavior, I had to supply not a new anatomy but new information.

Suppose, though, that parts of a salamander brain in front of the medulla really have no direct relationship to what a salamander does with a worm? Suppose feeding stations exist in the medulla or spinal cord (or left leg), awaiting only consciousness to ignite them? If this were true, the attack response on worms--the principal criterion in the study--would be irrelevant, and shuffle brain experiments would say very little about the holographic theory. A purist might have taken care of this issue at the outset.

"New experiments required," I scribbled in my notes. "Must have following features. Host: salamander minus brain anterior to medulla. Donor: try a vegetarian, maybe young Rana pipiens tadpole. But, first, make damn sure donor brain won't actively shut off salamander's attack on worms."

My working hunch was that the very young leopard frog tadpole would make a near-perfect donor. His taste for flies comes much later on in development. While he's little, he'll mimp-mouth algae from the flanks of a tubifex and harm nothing but a little vermigrade pride. Then, too, from experiments I'd carried out years before, I knew frog tissues wouldn't manifestly offend salamander rejection mechanisms, not to the extent that they would be destroyed. Thus, if grafted brains didn't perish in transit across the operating dish, they would become permanent fixtures in their new heads.

Whether a tadpole brain would or would not actively shut off worm-recognition programs in salamanders I had to settle experimentally before calling Punky into the game. Here, I transplanted tadpole brain parts but left varying amounts of host-salamander brain in place. These animals ate normally, thus showing that tadpole brain, per se, would not overrule existing attack programs. As I had guessed, it was like adding a zero to a string of integers as far as feeding was concerned.

Now the scene was ready for Punky, the first of his kind through the run. He would surrender his own cranial contents in front of the medulla to the entire brain of a frog. If his new brain restored consciousness but gave him a tadpole's attitude about worms, he'd vindicate the shuffle brain experiments.

For controls, I carried out identical operations but used other salamanders as donors. Also, to assure myself that frog tissue itself would not affect appetite, I inserted diced tadpole in the fins and body cavities of still other salamanders. This procedure had no effect on feeding. Moreover, I had a hunch that Punky would remain blind. So I removed eyes from other salamanders to get fresh data on feeding via sonar.

Punky awoke on the seventeenth day. Very quickly, he became one of the liveliest, most curious-acting animals in the lab. He did remain blind but his sonar more than compensated. A fresh worm dropped into his bowl soon brought him over. He'd nose around the worm for several minutes. He lacked the tadpole's sucker mouth. And I couldn't decide whether he wanted algae, or what. But he spent a lot of time with the worms. In the beginning, he had me watching him, wondering in a pool of clammy sweat if he'd uncork and devour the holographic theory in a single chomp. Yet, during three months, with a fresh worm in his bowl at all times, in more than 1,800 direct encounters, Punky never made so much as a single angry pass at a tubifex. Nor did any of his kind in the months that followed. The herbivorous brain had changed the worms' role in the paradigm. They were to play with now, not to ravage.

I kept Punky's group nourished by force-feeding them fresh fillets of salamander once a week. This meant the same thing had to be done with each and every control animal too. While the extra food did not blunt control appetites, the added work left me looking groggily toward pickling time when I could preserve the specimens on microscopic slides.

I routinely examine microscopic slides as a final ritual. But Punky's slides weren't routine. And on the very first section I brought into sharp focus, the truth formed a fully closed circle in the barrel of my microscope. His tadpole brain, indeed, had survived. It stood still in terms of development, but it was a nice, healthy organ. And from its hind end emerged a neural cable. The cable penetrated Punky's medulla, there to plunge new holographic ideas into his salamander readout, and into the deepest core of my own beliefs.

NO BLINKERS FOR THE BRAIN
Professor John Taylor

The brain is, by no means, an understood structure.

The brain presents one of the greatest challenges to our proper understanding of the world around us. It is being probed ever more deeply by the electron microscope and the micro-electrode. Yet in spite of an enormous amount of data being accumulated about the brain, there is no equivalent increase in our understanding of how it achieves what it does in controlling our behaviour. At times, the accumulation of facts about the brain even seems to be in danger of engulfing the research worker, making it increasingly difficult to bring order out of the gathering chaos.

Naturally, there is a plethora of theories as to how the brain functions, and a great deal of interest is currently being shown in the problems by workers from traditionally theory-oriented scientific disciplines, particularly chemistry and physics. The former of these subjects has a rightful place among the parties interested in brain research, because much important brain activity is controlled by chemical reactions inside living cells or at their membranes. Physics also has a reason for being involved when the electrical activity of nerve cells is under discussion, either in the laboratory when the activity is being measured experimentally or on the blackboard in models of nerve cell excitability.

It is the legitimacy of the further activities in brain research of chemists, physicists, and especially mathematicians, which is at present engendering a great deal of heat but not much light. These activities are all centred around attempts to explain brain processes following their own particular disciplines.

From *New Scientist*, September 13, 1973; used with permission.

The two disciplines of physics and mathematics, most noted for their modelling activity, have given the largest number of theories of the brain, and workers seem to be flooding into brain research from them to add their contributions as rapidly as possible.

To try to reach an assessment of the true nature of the situation, the first interdisciplinary meeting covering theory and experiment in brain research has just been held in Trieste, at the International Centre for Theoretical Physics. The setting was ideal, the International Centre for Theoretical Physics already covering various disciplines in physics and mathematics. Through set lectures and very open ended discussion groups, it was hoped to find out what the most suitable mathematical/physical approach is to the brain--if one exists. The participants covered the disciplines of chemistry, zoology, phychology, physics, and mathematics. Naturally enough no gross agreement was reached, though one very strong moral was drawn by the end of the meeting by a good proportion of the participants, and it is the title for this article.

The first problem which faced a number of participants was to learn what there was to explain--what the detailed sub-problems about the brain were. This led to the first real difficulty--that of the lack of gross variables similar to the thermodynamic ones describing the bulk properties of matter. Parameters are recognisable only in the case of very simple behaviour patterns, which restricts discussion either to a very limited range of responses or to very simple animals. Neither of these alternatives is at all helpful in discussing man.

This problem is one that is quite foreign to the sciences of matter, and also to mathematics, which requires quantifiable variables of some sort before it can even be used at all. From the talks and discussions it became clear that there are two quite distinct ways of approaching this situation. One is to assume that thermodynamic variables for the brain will be found sooner or later, and to set up general mathematical schemes to describe this situation. The difficulty with this is that it is almost impossible to make any predictions, because a dictionary translating the quantified variables of the theorists into specific quantities measurable by an experimentalist cannot be available. There are also mathematical approaches to the brain which may be so general as to be applicable to almost any brain. Automation theory and catastrophe theory are examples. These theories are general frameworks which apparently have yet to prove their value; they certainly make no predictions, nor seem within a light year of doing so.

This method of tackling a problem by side-stepping it and looking at a different, easier problem is dear to the hearts of mathematicians and it was they who in discussion fought strongly for this modus operandi in brain research. The physicists' natural bent is to look for the atoms of the system, and explain gross properties in terms of those atoms. This has proved extremely successful in physics so far, with the sequence: molecule→atom→nucleus plus electrons→protons, neutrons plus electrons→elementary particles still in the process of being extended to quarks or partons. The process of reversing the sequence has not proved easy, however.

The first atom of the nervous system is the nerve cell, and its wiring diagrams were beautifully described in certain cases. The unit controlling fine motor activity, the cerebellum, is especially well analysed, and possesses the most remarkable symmetry. Indeed, this was an example of a network of billions of neurons with an essential simplicity to it which reduces its analysis to only a few nerve cells. Very recent work on the cerebellum of the frog has shown how this organ controls very fast ballistic movements; a theory presenting it as a timing device was found seriously wanting, the unit of time involved being far too short for the movements being considered.

The other region of the brain considered in great detail was the cerebral cortex, considered to be man's crowning glory. Various theories for its functions as an "association machine" were considered, though not one was found completely satisfactory. None of them involved the feature detectors so important in vision, so they could only contain a limited version of the true picture. A dearth of known network structure was also seen here, though the lack of evident symmetry makes the cerebral cortex a much harder problem than the cerebellum.

Besides movement and intelligence, the other general feature of the brain of great interest is memory. One of the most exciting things at the meeting was the pictures of the detailed structures involved at the intermediate region between two nerve cells at which information is exchanged, called the synapse. There were evident little packets of chemicals which, released into the synapse from one cell, cause the next cell to become active. This was not new, but now fusion of these vesicles with the presynaptic membrane were revealed, and modification of this synaptic region with experience--unanaesthetised, anaesthetised, and near-death--were also clear. In fact there was now hope for a realistic search for a local theory of memory, experience producing long-lasting modifications of the ease with which one neuron acts on another.

To make such a dream of understanding memory come true, it was realised that it would be necessary to build back up to gross behavioural changes from local ones. Here the gap between the single nerve cell and the brain was met again. Various levels of modelling could be discerned to bridge this gap, from the rather descriptive approach (hand-waving, more or less), to a slightly more quantified attempt to assess information content by various means, to a much more full-blown use of the properties of single nerve cells.

None of these essays in going from the nerve cell to the nerve net proved successful, so that the relevant mathematical techniques, say, of information theory or the theory of equations, could not prove its worth. What has to be done before the brain can be properly understood is to bridge the nerve cell--nerve net gap completely, so that the methods used in all of the above models at the various levels would be required ultimately. In other words various branches of mathematics--such as systems theory, information theory, automation theory, and theory of equations--will all be needed. However the exact aspects required are not yet known.

The conclusion of all this was that we do not yet know the exact mathematics and physics relevant to explaining brain function; it would most likely be a suitable amalgam of the topics mentioned above with others still to come and possibly various aspects of electrical circuit theory (to describe each single nerve cell). In such a situation it is evidently unwise to plump heavily for one or other mode of description to the detriment of the others. To find the magical recipe for the right mixture of disciplines to give the key to brain function, we must keep as closely as will be guesswork with little chance of success.

Scientists interested in turning their thoughts to brain research should take note of this moral. It is well represented by the story of the two scientists, one theoretical and the other an experimentalist, who were walking along the street one night. One dropped a key, and they both began searching for it. However, the theorist walked over to a street lamp a little distance away and began searching underneath it. When asked why he replied "because I can see over here". This problem of looking where it is easiest, in spite of its irrelevance, is a particularly big one for physicists and mathematicians. There *is* no easily-quantifiable physical or mathematical scheme in which the brain can be embedded; the brain's equivalent of Newton's laws have not yet been discovered. They will only be so if all blinkers are removed and distant seductive street lights are ignored.

With the right attitude undoubtedly we will understand in the end; to coin a phrase, mind must always triumph over matter.

Professor John Taylor is head of the department of mathematics, King's College, London

A HEART-STOPPING, EYE-BULGING, WAVE-MAKING IDEA
Scot Morris

> We are beginning to realize that we can have more influence over our body functions than has been thought.

You would never guess that Swami Rama has been practicing yoga since the age of four and heads a monastery in Rishikesh. Now, in his mid-40s, he looks more like an Italian nobleman in his Nehru jacket and turtleneck sweater, and so he appeared very much out of place in Topeka, Kansas.

He was there for a short visit with Dr. Elmer Green, director of the Menninger Foundation's voluntary-controls project; and on the morning he was to leave, after completing an extensive series of tests in Green's laboratory, he decided that it was time to stop his heart.

"Are you sure you want to do this?" Green asked.

"Yes, yes," the swami reassured him. "There will be no problem. If you like, I'll sign papers releasing the Menninger Foundation of any responsibility."

"Oh, that shouldn't be necessary, but I'm worried about your health. You say you'll stop your heart by controlling your vagus nerve. But the vagus also innervates the stomach--you could get very sick. You had dinner at our house last night, you know."

"I know," said the swami, "and usually I would fast for two or three days before trying this. But I'm leaving Topeka today; I have a plane to catch. And besides, I want to see if I can do it." He paused and turned to Green. "You would like to see a man stop his heart, wouldn't you?"

"Well, yes, of course," said Green. "But I don't want anything to---"

Originally appeared in *Playboy* Magazine; copyright © 1972 by Playboy.

"I will do it, then," the swami interrupted. "Now, if my stomach were empty, I could safely stop my heart for three or four minutes, but I think I should limit the time today. How much will be sufficient for your measurement?"

"Ten seconds will be quite impressive," said Green, as he began to attach the EKG electrodes to the Indian's right ear and left hand.

Swami Rama had already shown remarkable control of his body in Green's laboratory. Three days before, he had caused the temperature of two spots on the palm of his right hand to differ by ten Fahrenheit degrees--the left side turned red, as if slapped by a ruler, while the right side turned ashen gray.

The electrodes were finally in place and everyone was ready for the heart-stopping test. Green stood by the swami's side, while Green's wife, Alyce, and a gaggle of technicians and professional colleagues sat in the control room watching the polygraph pens perform their strange spastic dance.

The swami made a few trial runs at speeding and slowing his heart, then said, "I am going to give a shock. Do not be alarmed."

Green naturally thought the swami meant he was going to give himself some kind of neurological shock. But the shock he was talking about was psychological. The swami was telling observers not to be concerned by what they were about to witness.

But no one, not even the swami, expected to see the mad pattern drawn on the EKG paper that day. Before the shock, his heart rate was smooth and even at 70 beats per minute; then, suddenly, in the space of one beat, it jumped to nearly 300 beats per minute. The polygraph pens reacted crazily, jerking up and down five times every second for a span of at least 17 seconds. When the swami arose, he was as surprised by the recording as everyone else. Yet he felt the test had certainly been successful: "When you stop the heart in this way it still trembles in there," he said, fluttering his hands to illustrate.

The swami's heart had experienced a sudden atrial flutter, a dangerous cardiac condition that usually renders a patient unconscious, a condition in which the heart vibrates so fast that blood does not fill the chambers properly, the valves fail to work as they're supposed to and no blood is pumped to the body.

The swami had used this technique many times to stop his pulse for the benefit of skeptical doctors, but he didn't realize until he saw the results of his EKG that in order to do it, he had forced his heart to flutter so fast that it couldn't function properly.

The theoretical impact of Swami Rama's demonstration was significant, not so much because of what he did--stopping his pulse for 17 seconds--nor how he did it; the demonstration was really important because it took place <u>when</u> it did. Actually, the same study could have been performed 20 years ago; equipment was available and yogis have claimed for centuries to be able to stop their pulse. But it was done in 1970 and, ironically, was much more impressive than if it had occurred five years earlier.

Why? Because in 1965, and for centuries before that, it was an accepted medical fact that the body's muscles were of two distinct types--voluntary and involuntary. Voluntary muscles, such as those in the arm, are triggered by the central nervous system, so one can move them by merely deciding to do so. But the muscles of the heart and other internal organs, it was believed, were entirely different. They were thought to be under the sole control of the autonomic nervous system, which, as the name implied, was autonomous: It acted independent of consciousness, automatically regulating the body's involuntary, visceral responses.

Therefore, in 1965, a traditional scientist would have called the heart-stopping demonstration a trick. "Anyone can change his heart rate by simply varying his breathing pattern," he would have said. "And we know it is possible to slow the heart drastically by tightening certain voluntary muscles that squeeze against adjacent blood vessels and block the flow of blood feeding the heart—the so-called Valsalva's maneuver. But this shows only that the swami has developed fine control of his <u>voluntary</u> muscles. He cannot actually control his heart directly, because the heart is an involuntary organ. The swami's feat is no more significant than if he had claimed voluntary control of his salivary glands, and then demonstrated it by voluntarily putting a lemon in his mouth."

Then, in 1966, Neal Miller and his colleagues at Yale destroyed the traditional argument with one ingenious experiment using rats injected with curare, a drug that blocks all the central-nervous-system outputs to voluntary muscles. (Curare is the drug with which South American tribes tip their poison darts. A victim feels the pain of the dart and remains conscious, with his brain and internal organs functioning normally, but he lies limp and helpless, unable to make any voluntary movements. Since his breathing muscles are among those silenced, he quickly suffocates. To prevent this unhappy event in his curarized rats, Miller kept them breathing with an artificial respirator.)

Miller's experiment showed that a rat with all its voluntary muscles blocked could still learn to raise or lower its heart rate in order to get the reward of a brief electrical charge to the "pleasure center" of its brain.

This one experiment proved that the heart was not involuntary and suggested that perhaps the distinction between voluntary and involuntary body organs should be thrown out the window.

Soon after it was known that rats could learn to control their hearts, many researchers reported that undrugged human beings could do the same thing. Psychologist Bernard Engel and his colleagues at the Gerontology Research Center in Baltimore have trained cardiac patients to control their own premature ventricular contractions, the irregular heartbeats associated with an increased probability of sudden death. Half of Engel's patients learned their lessons well enough to maintain good control at home and in later follow-up laboratory sessions.

Out of the knowledge that many bodily responses thought to be involuntary can in fact be controlled has grown the technique called biofeedback. All learning requires feedback. If you are learning to shoot a bow and arrow, you must be able to see the arrow and the target and know the results of your shot. If you are learning to speak, you must be able to hear. A child who is hard of hearing speaks poorly; given a hearing aid, he may learn to speak as well as a normal child. Biofeedback is simply feedback applied to biological matters. It acts like a hearing aid. It "turns up the volume" on internal body signals that are normally too faint to be heard.

By its very nature, biofeedback implies the use of machines. A biofeedback machine is any device that makes a person more aware of an internal bodily function than he normally would be and that the person uses in an attempt to control the function voluntarily. A stethoscope can be a biofeedback machine if a person uses it to observe his own heartbeat, to learn what it feels like to have his heart rate go up or down and to try to bring his heart under voluntary control. The stethoscope can be improved upon electronically, of course. The heartbeat might be amplified through speakers, making it possible to detect smaller changes. Or it might be connected to a visual display, so that a subject can look at a dial that **tells** him what his heart rate is at any moment. Or the signal might go to a computer that turns on a light whenever the heart rate is, for example, two percent faster than it was the previous minute. Obviously, the sophistication and expense of biofeedback machines can vary tremendously.

Theoretically, if a person can be made sufficiently aware of any bodily process under neural control, he can learn to control it voluntarily. This includes almost everything imaginable--heart rate, pupil size, brain waves, the secretion of hormones and digestive juices, even the activity of individual nerve cells. It also includes abnormal processes such as headaches, insomnia and systemic diseases. This is the theory, and the promise, of biofeedback.

Is it all theory? Does it really work? Although biofeedback is still in its infancy, enough research has been done so that it's fair to ask how well the early promises have been fulfilled. The answer is complicated and varied, as varied as the special areas of biofeedback that are now being investigated. In some areas the findings have been dramatic and exciting, almost spectacular. In others, the results have been less successful than we might have hoped.

Some of biofeedback's greatest success has been in relieving muscle tension. With special machinery that registers muscle activity with a tone or electronic clicks, it's possible to learn to relax a muscle completely or to isolate a single motor neuron and learn to "fire" it at will, making it sound, through amplified feedback, like a drum roll.

Variations on the muscle-feedback technique have already been used to partially rehabilitate stroke victims, to teach asthmatic patients how to breathe properly and to eliminate subvocal speech in slow readers--the silent mouthing of words that keeps people from learning to read any faster than speech speed (about 150 words per minute).

New York psychologist Erik Peper used muscle feedback, converted into electronic clicks, to cure a man's facial tic that was so pronounced at the beginning of treatment that the recording electrodes wouldn't stick to his head. The man described his successful therapy as a revelation: "I kept saying to myself, 'Relax...relax...' then I concentrated on the clicks. At that moment, a whole new world was brought before me. I felt a deep warmth... For once I controlled my body movements, they didn't control me... I told myself, 'Don't tick--and I didn't, and I cried for happiness."

In an even more dramatic use of the biofeedback technique, University of Wisconsin psychologist Peter Lang saved a nine-month-old infant's life. The baby weighed only 12 pounds and was being fed through a stomach pump because he vomited everything he ate. He was not expected to live. Lang measured the muscle activity along the baby's alimentary canal and gave him brief electric shocks on the leg whenever his esophagus started to back up. "After only a few meals with this therapy, the infant ceased to vomit," said Lang. "He is now a healthy toddler."

On the other hand, heart-rate-control experiments have been somewhat disappointing. The <u>have</u> proved that some control is possible, and this has now been confirmed many times at Harvard and the universities of Wisconsin and Tennessee. But the changes achieved so far are still rather small and the best subjects have been young volunteers, not the more elderly cardiac patients who really need this treatment.

Blood-pressure research has also been inconclusive. So far, only a few researchers claim to have helped patients with essential hypertension, one of the most common forms of high blood pressure. Neal Miller, now at Rockefeller University, claims temporary success with one patient, and Albert Ax at Detroit's Lafayette Clinic says he succeeded with one hypertense patient and failed with another. The Harvard team of Herbert Benson and David Shapiro has had good results with seven patients, but only in a laboratory setting. Still, other Harvard researchers have demonstrated that some people can learn to manipulate blood pressure and heart rate independently--making one go up while holding the other steady.

It is obvious that a person could learn to control his own body with biofeedback to the point of potential physical harm. According to a <u>Wall Street Journal</u>

report, one enterprising specialist is teaching men how to fail their military physicals by raising their blood pressure to pathological levels.

At the Menninger Foundation, researchers are using biofeedback to cure migraine headaches. The same emotional reactions that cause the main artery of the head and face to swell, resulting in a painful migraine, also cause blood vessels to constrict in the hands, making them cold. This concurrence led Joseph Sargent to a remarkable kind of biofeedback therapy: He treats migraine headaches by teaching patients to raise the skin temperature of their hands. The therapy has now been used with over 70 patients, and about 80 percent of them have shown definite improvement; they use less medication and their headaches are fewer, shorter or milder. One woman patient later used the technique to warm her feet voluntarily, relieving a chronic source of her insomnia.

Many factors besides age work to determine how well a person responds to biofeedback training. Peter Lang argues that there are "autonomic athletes" and "autonomic duffers." Other research suggests as much: It notes that musicians, artists and athletes are especially adept at controlling their brain waves. Another interesting, and logical, finding from University of Tennessee psychologist Jasper Brener is that actors may be able to learn heart-rate and sweat-gland control more readily than the rest of us. Taking it one step further, Bob Stern at Penn State reports that Method actors, who try to experience each emotion they portray, can learn to control galvanic skin response (an activity related to sweating in the palms) much faster than can non-Method actors.

Of all the frontiers being explored with biofeedback, one has received more publicity than all the others combined: the alpha wave. The market now carries perhaps 30 brands of low-cost "alpha machines," which beep or buzz or light up to signal that the wearer's brain waves have entered the mysterious alpha state.

One firm capitalizing on the craze says that by learning alpha control "you will relax physically and mentally...control fears...find peace and vitality...solve problems while you sleep...create an amazing memory bank in your own mind...learn to develop extrasensory perception..."

Such advertising, with an assist from overzealous newspaper and magazine writers, has led to an almost unanimous misunderstanding of what alpha is.

Perhaps the most surprising, and nearly accurate, description of the alpha state was given by Swami Rama, during brain-wave tests with Dr. Green. After he had explored his inner states of consciousness with the aid of Green's feedback machine--which sounds a different tone for each brain-wave frequency, including alpha --the swami arose and said, "I have news for you. Alpha isn't anything. It is literally nothing."

In truth, alpha is something that occurs when you <u>feel</u> nothing. It is a brain wave with a frequency between about eight and 13 cycles per second (cps). Brain waves are the constant undulations in electrical activity that occur in any living brain. These waves are picked up by sensitive electrodes at the scalp and transferred to an electroencephalogram (EEG). Brain waves are a rough measure of mental activity: They are fastest during active, attentive thought and slowest in deep sleep. Generally, stimulants such as caffeine, tobacco and amphetamines speed up the predominant brain waves; while alcohol, morphine, marijuana and decreased blood sugar tend to slow them down. The alpha state is typically described as relaxed, pleasant, detached from reality.

Alpha is a steady, clean rhythm that stands out from the apparently random squiggles on a typical EEG record like a 4/4 drumbeat in a composition by John Cage. Faster activity (about 13 cps) is called beta. It signifies that the brain is active--worrying, perceiving, deciding or attending. Waves slower than alpha (below eight cps) are called theta and delta. The latter two usually occur only in drowsiness or sleep.

If an average adult is relaxed but not sleepy, if his eyes are closed and he's thinking of nothing in particular, he is probably "in alpha." (I specify average adult because infants do not have alpha waves, nor do about ten percent of adults, who are otherwise perfectly normal.) If you were to interrupt an EEG subject during his alpha state and ask, "What were you just thinking about?" he would probably answer, "Nothing." But as soon as you ask the question, the alpha waves would disappear and be momentarily replaced by faster beta waves. This response is called alpha blocking, and a person in the alpha state experiences it whenever something catches his attention.

Albert Einstein was said to be able to solve mathematical problems in his head while his alpha waves chugged merrily on. But if someone gave him an unfamiliar problem that required conscious thought, his alpha state became blocked like anyone else's. Einstein was so familiar with the world of formulas and figures that mathematical thinking came automatically, by second nature.

Alpha waves do not indicate total inactivity. They may occur while a person is actively doing something--whistling, peeling potatoes, even driving a car; but at these moments he is behaving automatically and is paying no attention to what he's doing.

So, in a sense, alpha <u>isn't</u> anything, as the swami said, and it is especially not most of the things people seem to think it is. Alpha is not a state of mind for spiritual enlightenment, creative insight or intuitive wisdom; it does not bring telepathic or psychedelic visions; it does not cure diseases.

The first popular report that people could be trained to control their own alpha waves appeared in a 1968 <u>Psychology Today</u> article by Joe Kamiya, now at San Francisco's Langley Porter Neuropsychiatric Institute. Experimenters rang a bell at random and asked patients hooked to an EEG machine to guess whether or not they were in alpha. By guessing correctly, patients eventually learned to produce alpha waves. Of the people who did, 80 to 90 percent could learn to control them to some extent.

The early stages of alpha-feedback training can be quite frustrating. As a subject relaxes and his mind goes blank, his brain waves start to enter the alpha range. Then the tone comes on and, of course, the subject notices it. But this disturbs his alpha state, so the tone stops. At this learning stage, the feedback seems to defeat its own purpose. After several trials, however, the tone no longer disrupts the subject's relaxed state; it fades into the background, like a refrigerator's hum, and one learns to hear it without listening to it.

Kamiya's discovery was significant and far-reaching, but it took on extra drama and intrigue when he pointed out that the mental discipline necessary to maintain an alpha state was connected with Zen and yoga meditation. Prior research at Tokyo University had shown that Zen meditators increased their alpha as they began meditation; and the more experienced Zen masters showed the most pronounced increase.

So Kamiya tried his alpha-feedback technique on seven practiced Zen meditators and found, not surprisingly, that they learned alpha control much faster than his other subjects did.

Yoga masters, too, have prominent alpha waves. At New Delhi's All-India Institute of Medical Sciences, students with naturally high alpha levels do better in their yoga training than students who have lower alpha levels to start with. (A high alpha level does not necessarily mean that the alpha wave is strong but that it is present much of the time, and in a large proportion of the brain.)

These findings inspired writers to report, a couple of years ago, that scientists had discovered the brain state corresponding to <u>satori</u>, and that we could now reach that state with a machine.

Such sensational speculations are easy to criticize, but, as a matter of fact, they may not be far from the truth. There are many striking similarities between the alpha state, as scientists understand it, and various meditative states as they have been described by mystics.

For one thing, their basic methods and aims are similar. Ramana Maharshi has said that the aim of yoga is "the cessation of mental activities," a phrase that could just as easily describe the aim of alpha training. In order to reach alpha, one must block all distractions. To accomplish this difficult task, some people center their attention on a single focus and come back to it whenever they find that their minds have wandered. This centering soon becomes automatic and effortless and alpha waves begin to appear.

Similarly, all forms of meditation use centering devices designed to reduce distractions. A beginning Zen student is told to concentrate on his breathing and to return his attention to it when his mind wanders. A Zen master will dwell on an irrational riddle, or koan, such as "What is the sound of one hand clapping?" or "What did I look like before my parents met?" Yogis may concentrate on a visual center--a mandala--or on a point at the center of the body. And in transcendental meditation, everyone has his own mantra, a phrase or sound (such as "Om") that he repeats over and over during meditation. Borrowing from the vocabulary, the instruction manual for a popular home alpha machine states that "the tone of the Toomim Alpha Pacer may be used as an effective mantra."

Considered in this light, all the stereotyped practices of the meditator--sitting in a prescribed posture, contemplating the navel, burning incense and renouncing all worldly desires--act to reduce potential disturbances and therefore to increase the likelihood of alpha waves.

Although a distracting stimulus will block alpha, a constant, unchanging one will facilitate it. This happens through a mechanism called habituation. We tune out anything that is familiar and repetitious. When someone is in the alpha state while a bell sounds, his alpha will block, giving way to faster, low-voltage beta waves. Alpha waves then return a few seconds later. When the bell sounds a second time, alpha blocks again but resumes sooner. Each time the bell rings, alpha blocks less and less, until eventually it is not disturbed at all.

Undeniably, habituation has its benefits--one can read this article, for example, without being distracted by the shape of the letters on the page, the feel of the clothes on one's back or the sounds of traffic outside. But habituation also causes a certain sensory impoverishment; it restricts awareness and makes one miss the rich, changing subtleties of the sensory environment.

Meditators can apparently override this automatic tuning-out mechanism. The Zen master responds to the 20th bell ring with the same attention he paid to the first. In technical terms, his alpha block does not habituate. This seems to corroborate Zen Buddhists' frequent claims that they reach a state of "perpetual here and now" with meditation and can "look at the world with a fresh eye, as if seeing it for the first time." Research on yogis shows that alpha is not blocked at all, even if lights flash or gongs sound or the skin is touched with a hot test tube.

Could the rest of us learn to reach this painless yogic oblivion with biofeedback? It may well be possible, and Erik Peper has done some promising research at the College of Mount St. Vincent indicating that if a person can learn to produce a strong, stable alpha rhythm he may be able to withstand a painful stimulus, such as a dentist's drill.

What's amazing, finally, is not so much that we can obtain fine physiological control with biofeedback devices but that Zen and yoga disciples have been able to do it for centuries <u>without</u> machines. The only explanation is that they are able to quiet their conscious minds so completely as to detect faint internal signals the rest of us can never hear without electronic amplification.

Neal Miller found an analogous effect in his rats. Normal animals could not learn to control their heart rates nearly as well as could rats injected with curare. The paralysis in the drugged rats apparently eliminated their muscular distractions and allowed them to attend to the subtle signlas coming from their hearts. (Miller reports that in the past two years, researchers have been unable to teach rats the same degree of heart-rate control that was obtained in earlier experiments. He is now investigating the possibility that in the past few years there may have been changes in the quality of curare available to researchers or in the strain of rats supplied by breeding houses. At any rate, the new difficulties do not change the earlier conclusion that the heart is not an involuntary muscle.)

I once asked Dr. Barbara Brown, chief of the experiential-physiology lab at the Sepulveda, California, VA hospital, whether there were noticeable personality differences between people with naturally high and naturally low alpha levels. She paused a few seconds, then asked, "Do you know what your alpha-wave level is?"

"Not really," I said. "The machines I've tried so far weren't very good."

"Well, then I can answer the question without hurting your feelings," she laughed. "In general, people who tend toward a lot of alpha in their EEGs are dull, uninteresting, unimaginative, hard-working, plugging-along, ordinary people. And that's the truth. One finds this in everyday lab experience. All you have to do is look at an EEG record and you can tell what kind of personality the person has."

This doesn't mean that a person who trains with an alpha machine will definitely become the dullard Dr. Brown describes--and many researchers would consider her generalizations exaggerated--but there is no guarantee that he won't, either.

Certainly, many people find the alpha experience quite pleasant, and they understandably wish to increase the amount of time they are in alpha. But there are many ways to do this. Just sitting quietly and thinking of nothing will result in a significant increase in alpha. Defocusing the eyes will do it, too, as will training in deep muscle relaxation. Many people can increase their alpha waves by simply rolling their eyes upward as far as possible. (This, by the way, is the position assumed by some yoga meditators when concentrating on the <u>ajana</u>, or "third eye.") "If all you want to do is have lots of alpha," said Dr. Brown sarcastically, "you can stuff yourself full of marijuana, or be a chronic LSD user, a heroin addict, a depressive schizophrenic, or have specific kinds of behavioral problems."

You can also use an alpha-feedback machine. Unfortunately, on many cheaper machines the tone will respond not only to alpha but also to theta or muscle tension. Dr. Brown says that one man who had been training on a home alpha machine recently volunteered to demonstrate his abilities on her laboratory EEG machine. When the man went into the state he called his alpha, nothing came through the EEG but exaggerated eye movements. The man's machine had been responding to his muscle potential, and by trying to "keep the tone on," he had taught himself a pronounced eye tremor.

Sometimes the consequences can be more serious. Some of the brain waves associated with epileptic seizures will trigger many alpha machines. A potentially epileptic person who tries to keep the tone on may be horribly surprised if he succeeds.

After my conversation with Dr. Brown, I tried alpha feedback on the sophisticated research machine in her laboratory. It told me not only when alpha was present but also how strong it was and where in the brain it was coming from. I loved it. After the extended session I felt very relaxed, very good. I would recommend the experience to anyone.

But while the session made me feel good, _being_ in alpha, as such, was not necessarily the source of the good feelings. First of all, there was an exhilarating feeling of self-mastery. Realizing that I could control my own brain waves was thrilling. The pleasure was in doing it, in turning alpha on and off at will. But staying in alpha was boring. I don't like enforced mental inactivity. Others do, apparently, and for them an alpha machine can be a splendid enforcer.

Also, I'm sure the power of suggestion colored my reactions. We are told that the alpha state is quiet, tranquil, pleasant. So whenever the feedback signal came on, I thought, "Ah, it's working. I'm relaxing. The machine says so." And so I relaxed. The power of suggestion plays an essential part in the alpha experience. Research at the Bedford, Massachusetts, VA hospital has shown that before a person will state that an alpha-feedback tone is associated with pleasant feelings, two conditions must be present: (1) The tone must really be linked to the person's alpha and (2) he must be led to expect that the tone will be correlated with a pleasant inner state. The absence of either of these conditions has resulted in subjects' reporting that the experience was not pleasant.

Stating that the alpha experience makes you relax is misleading. It implies that a subject somehow makes his brain cells fire at eight to 13 cps and, as a result, he relaxes. Actually, the reverse is more accurately the case. The alpha experience _is_ relaxing, but it is not that alpha waves as such cause you to relax. Rather, you must relax in order to achieve the alpha state. Since a session with an alpha machine is really nothing more than a period of monitored relaxation, people emerging from the experience can feel no way _but_ relaxed.

The confusion comes from the understandable tendency to summarize biofeedback's advances in quick and easy terms. We say that a person has learned voluntary control of his heart rate or skin temperature, which leaves the impression that he can somehow flex his ventricles like biceps, open and close his blood vessels at will and direct his blood flow. Actually, he does not directly learn to control his heart or his blood vessels or his alpha waves; he learns only to keep the tone on. He does this, first, by trying to detect some internal event—a thought, an image, a feeling—that seems to correlate with the tone, and then by concentrating on that event to reproduce it as often as possible. At this point, when one knows _what it feels like_ to slow his heart, lower his blood pressure or enter the alpha state, the feedback machine is no longer needed and the person can continue his physiological control unaided.

It is unfortunate that so much attention has been focused on ways to achieve the alpha state, because some of the most exciting biofeedback research deals with methods of _avoiding_ it.

Tom Mulholland, past chairman of the Biofeedback Research Society, has suggested that automobile drivers and heavy-machinery operators might wear an "alertometer," which would sound an alarm whenever their brains drifted into alpha. In a more immediate application, Mulholland is using this approach with children who have learning difficulties often accompanied by short attention spans. He has developed a teaching machine with a slide projector that throws an image only when the student is in beta, meaning that his attention is strongly focused. When the child's brain waves slip into alpha, the machine stops until he is again attentive and ready to continue.

Advertising agencies are using these principles to test the effectiveness of their commercials: Any ad that doesn't block alpha is obviously ineffective.

A few years ago, Barry Sterman, chief of the neuropsychology lab at the Sepulveda VA hospital, noticed that when cats stand very still, the waves in the sensorimotor part of their brains are predominantly 12-14 cps. When he trained cats to produce this frequency to get a food reward, they became quiet and still, even when injected with a drug that ordinarily produces convulsive seizures. Since then, Sterman has trained three epileptic patients to generate this 12-14 cps wave and says that their seizures diminished markedly and their day-to-day EEG patterns have become more normal. He cautions that controlled studies must yet be done, but at this point the research is very promising; this is the first report of a lasting change in a person's EEG as a result of biofeedback.

Other researchers are studying theta waves, the four-to-eight-cps frequencies that seem to be associated with drowsiness, reverie, recall, creativity and the experience of "expanded consciousness" associated with LSD trips. As Elmer Green describes it: "There is a door to your inner self which is locked under normal circumstances. LSD is one of the keys to this door, but it throws you directly into an uncharted, unexplored world, unable to control how far in you go and how fast you get there. Thus, the 'bum trips.' With feedback, you enter slowly, feel your way around and proceed a step at a time. And you can turn around and come back out any time you so desire."

Green and his wife are studying the relationship of theta waves to hypnagogic imagery, which is defined as "pictures or words that are not consciously generated or manipulated but which spring into the mind full-blown." Many people experience these images just as they are falling asleep; when these images occur, their EEGs are likely to register theta waves.

August Kekulé, a German chemist, once dreamed of atoms undulating in a snakelike chain. Suddenly, one of the snakes held its own tail in its mouth. The image led Kekulé to postulate the existence of the benzene ring, which has been called the most brilliant piece of prediction in organic chemistry.

Niels Bohr's conception of the structure of the atom, the invention of lead shot for shotguns and Elias Howe's sewing machine--all came to their creators as hypnagogic images.

Robert Louis Stevenson regularly used this state of mind to generate tales such as Dr. Jekyll and Mr. Hyde. He would drift into reverie and command "the brownies" of his mind to furnish him with a story while he slept. He could even return to this state later to change an unsatisfactory ending.

Of course, we don't know the nature of these men's brain waves at the moment of their creative insights, but from what we have learned about hypnagogic imagery, it's probable that they were theta waves.

It's been very difficult to study the theta state and hypnagogic imagery in the past, because we don't stay in theta very long; we usually pass through it briefly on the way to sleep.

The Greens have developed an ingenious instrument to overcome this problem. Their feedback machine has been devised so that each brain-wave frequency has its own musical tone: The fast beta waves sound like a piccolo, alpha sounds like a flute, theta like an oboe and delta like a bassoon. (Green says that in grant proposals he sometimes claims he will try to train subjects to play The Star-Spangled Banner with this brain-wave orchestra, in hope that Government granting agencies will be encouraged to release more funds for research.) The machine will even record the information from the left and right sides of the brain and feed it back in stereo--the music of the hemispheres. In theta training, the machine feeds back only the oboe sound of theta.

To keep a subject from passing through theta into sleep, Green added an imaginative secondary feedback trigger. Whenever alpha disappears for 30 seconds, a bell automatically sounds to rouse the subject. In this way, he can remain in a predominantly theta state for extended periods of time. He can observe his unconscious images, surface to consciousness and report on them, and then return to theta.

"Alpha is normally being aware and theta is being unaware," Green says. "And if you can learn to stay in the phase where both waves are present, you may become aware of normally unconscious material. Apparently a certain amount of alpha must be present so that a person can bring the hypnagogic images over to consciousness."

Green's studies with Swami Rama shed some light on this point. After several periods of practice with the feedback machine, the swami announced that he would produce delta waves in the laboratory. No adult had ever shown slow delta waves while awake; they are present only in deep sleep. Yet the swami was able to produce delta waves during a prearranged period of 25 minutes, then waken and report accurately on what had happened in the room while he was asleep. The relevant point is that the swami's record never reached above 40 percent delta; the rest of the time was spent about equally in the other brain waves. Green speculates that it's necessary to retain some alpha in order to be aware of what is happening on the unconscious stage, but not so much that unconscious levels are lost.

Charles Tart, a psychologist at the University of California at Davis, suggests an easy way to study hypnagogic imagery at home: "Lie flat on your back, as if going to sleep, but keep your arm in a vertical position, balanced on the elbow, so that it stays up with a minimum of effort. You can slip fairly far into the hypnagogic state this way, getting material, but as you go further, muscle tone suddenly decreases, your arm falls and you awaken immediately."

Will people trained by theta feedback become more creative and generate great inventions or Nobel Prize-winning insights? Green is reluctant to say while his research is still going on. And Barbara Brown, who was delightfully candid about much of her biofeedback work, was circuitous and noncommittal when asked about her ongoing theta-training studies.

Most researchers have become restrained when speaking to reporters. They have seen too many impossible headlines ("TURN ON WITH ALPHA"; "THINK YOUR HEADACHES AWAY") that make biofeedback sound like vitamin C or copper bracelets.

Of all the major researchers, none has received more publicity than Joe Kamiya, the man who first demonstrated voluntary control of alpha waves. For a while he seemed to revel in the publicity and was free with his speculations to reporters. He talked about eventually curing nearly every bodily illness through biofeedback. He no longer speaks so freely. "There's been too much publicity for me," he said, "and I feel as though my work is not yet really ready for presentation." He is concerned that his colleagues will regard him as a publicity seeker: "I, more than any of the others in the field, have been behind in my [professional] publications." Now Kamiya is even discounting the potential of alpha control. "By itself," he has said, "alpha feedback is not a useful therapeutic tool at this time." And Neal Miller adds, "Learning the skills of alpha control may have no more significance than learning to move one's ears."

Some traditional scientists resent the fact that researchers who have lately hopped onto the biofeedback band wagon are concentrating on the most spectacular areas of study rather than helping to work out the background parameters that will define biofeedback's scope and limits. As Peter Lang put it: "A lot of people are going for the throat before they know where it is."

Finally, Miller fears that the public will expect too much too soon: "This exaggerated optimism may lead to inevitable disillusionment which will prevent the hard work that has to be done to see what therapeutic value, if any, there is in this approach."

So, publicly, biofeedback researchers have become cautious, underselling their own discoveries. But they certainly haven't given up their experiments and hundreds of other scientists have joined them. Between 1969 and 1972, the Biofeedback Research Society grew from about 20 members to over 300.

There's little wonder that biofeedback has been sensationalized by the press: It *is* sensational, one of the most exciting new fields in science. We can now explore objectively those obscure, dimly comprehended states of consciousness that have been talked about for centuries by yogis and mystics, and more recently by hypnotists and acidheads.

The possibilities in medicine are overwhelming. Once one understands the basics of biofeedback, the speculations seem to pour forth. Can we control heart rate and blood pressure? Then perhaps we can prevent deaths from heart disease. Can we control brain waves? Then perhaps we can control epilepsy, personality disorders and learning deficiencies. Can we overcome the alpha-blocking response? Then we may learn to ignore pain. Can we control blood flow? Then we may be able to teach a man to have an erection at will, or to kill a cancerous growth by starving it of its needed blood supply. Someday, perhaps, the ancient dictum will be changed to "Patient, heal thyself."

And, potentially most significant, many physicians estimate that about 80 percent of human ailments are at least partly psychosomatic. If the mind can make the body sick, perhaps it can also make it healthy. This concept means not only that abnormal disorders can be treated but that everyone's general health can be raised to new levels. So someday we may all be taking courses in psychosomatic health. If so, biofeedback researchers are already building the teaching machines that will be used in the classroom.

GADGET KEEPS THE TUMMY TIGHT

A device that toots when your tummy hangs out has been invented by a nuclear physicist in Chicago. Belt with a small plastic box of space age electronics sounds a buzzer when your abdominal muscles relax. Invented by Dr. William Bertram of Chicago, stomach reducing belt works on principles of isometrics and Pavlovian response.

OUT ON A LYMPH
with BURT BACTERIA

HEY ALL YOU BODY FREAKS. DO YOU LIE AWAKE NIGHTS WONDERING WHY **CHICKEN POX** ONLY STRIKES ONCE? HAVE YOU SPENT YOUR LEISURE HOURS MUSING ABOUT HOW **VACCINATIONS** PROTECT YOU FROM THE ONSLAUGHT OF **MICROBIAL MENACES**? DO YOU STOP PEOPLE ON THE STREET TO COMPARE YOUR ANNUAL **COLD-CATCHING RATES**? OF COURSE YOU DON'T — **AND WHY NOT?** BECAUSE YOU REALLY DON'T CARE, DO YOU — FESS UP! YOU'RE WALKING AROUND IN A REGULAR SCIENTIFIC MARVEL AND YOU GIVE MORE THOUGHT TO WHICH OF THE THIRTY-ONE FLAVORS TO CHOOSE THAN TO THE FACT THAT YOU'RE BOMBARDED DAILY WITH MILLIONS OF DISEASE CAUSING ORGANISMS AND RARELY DO THEY GET THE BETTER OF YOU. WELL, BE THE FIRST ON YOUR BLOCK TO BE GRATEFUL FOR THE POWERS OF **IMMUNITY**. THE MAKERS OF **GO TO HEALTH** PROUDLY PRESENTS AN **ADVENTURE IN IMMUNELAND**...

STARRING —

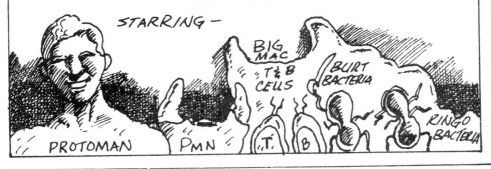

PROTOMAN · PMN · T & B CELLS (BIG MAC) · BURT BACTERIA · RINGO BACTERIA

Art by Karl Nicholason, from *Go To Health*, copyright © 1972 by Communications Research Machines Inc. Used by permission.

BIG MAC is a regular bag of tricks.... Inside his amoeba-like body are bags of enzymes called LYSOZOMES that fuse with the ingested bacteria. Now, it would seem reasonable to assume that our story should end here but as tough as BIG MAC is he is just a link in the chain of IMMUNITY....

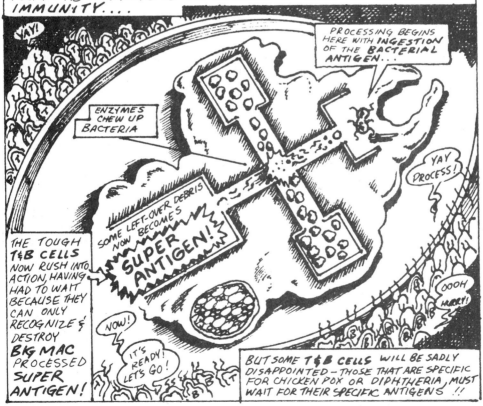

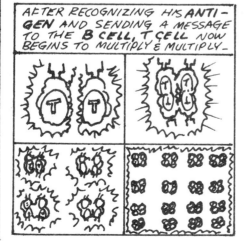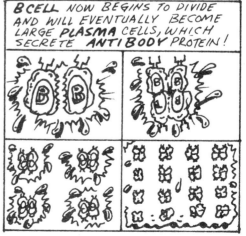

STALKING THE WILD CROWN GALL
Wallace Cloud

>Despite prolonged research, cancer remains an inscrutable malfunction of our bodies.

Virus-hunting has become a preoccupation of many cancer researchers. Although no virus has been proved to cause cancer in human beings, laboratory work on animal cancers heavily implicates these submicroscopic "particles," which hover on the borderline separating large chemical molecules with complex structures from living entities. Much is now known about how these intracellular parasites intervene in the genetic apparatus of their host cells--but much remains to be learned. Thus, there was standing room only at a particular session of the recent American Institute of Biological Sciences meeting in Amherst, Massachusetts. Dr. Christine F. Pootjes, a microbiologist from Penn State speaking on crown gall disease, reportedly had findings which might indicate whether the mysterious tumor-inducing principle in this plant cancer was a bacteriophage--the specialized viral form that invades and multiplies in bacterial cells.

Can study of a plant cancer yield clues that might lead to cures for human cancers? Evidently, the decision-makers at the National Cancer Institute think so, since that agency has funded about a dozen research projects in this field. According to Dr. Philip Stansly, NCI program director for cancer biology, crown gall is an "intriguing" model system "that allows certain kinds of manipulation and therefore provides the possibility of certain kinds of insight that you cannot get with the conventional animal models" widely used in the study of cancer. He points particularly to the work of Dr. Armin C. Braun, head of the Laboratory of

From *The Sciences*, September 1973. Copyright © 1973 by The New York Academy of Sciences. Used with permission.

Plant Biology at Rockefeller University, and probably the most widely respected crown gall researcher.

SURE-FIRE MALIGNANCIES

Like most cancers, crown gall is a killer: it almost wiped out the raspberry farms of Wisconsin 30 years ago. But this disease is unique among cancers in that it is known to be transmitted by bacteria. When and where human cancer will strike is unpredictable, but in crown gall "the [bacterial] transformation efficiency [from health to malignancy] is just fantastic," says Dr. Braun. "You can get literally 100 per cent transformation if you inoculate a susceptible plant with these bacteria." Inoculation consists simply of making a plant wound and introducing the bacterium Agrobacterium tumefaciens or certain other varieties of agrobacter, which induce tumors closely related to crown gall.

That what happens is the autonomous growth of true cancer tissue has been shown in several ways. The bacteria can be killed by placing the inoculated plant in a room heated to 115 degrees F, but the tumor tissue continues to proliferate. The tumors also metastasize—in other parts of the plant, new tumors appear which are generally bacteria-free. Such tumor tissue can be cultured in vitro on a very simple medium providing only sugar, a few minerals, and three vitamins; this means that the tumor cells have become independent of the auxins and other growth hormones usually required by plant tissues.

There is considerable disagreement over the "tumor inducing principle" being hunted by many of the 30-odd U.S. and European research groups working on crown gall. Is the TIP bacterial DNA (deoxyribonucleic acid, the genetic material that conveys biological replication instructions in molecular code), or a fragment of bacterial DNA which may be replicated by tumor cells? Or is it viral DNA? Researchers know the agrobacter strains are normally infected by a strain of bacteriophages. I asked the opinion of Dr. James A. Lippincott of Northwestern University, who spoke at the AIBS meeting. He laughed and said, "I think it's a black box."

VIRUSES IN HIDING

The bacteriophages are "little fellows with polyhedral heads and long curly tails," said Dr. Pootjes in the course of her eagerly awaited talk. But are these viruses part of the crown gall cancer system? The phages could provide the DNA needed to transform normal cells into neoplastic cells, she said; they have been isolated from bacterifree tumor tissue, although that fact doesn't indicate their role in transformation, if any. Some investigators have reported that tumors could be induced by virus-free strains of bacteria—so-called cured strains. However, Dr. Pootjes reported, when such "cured" bacteria were stressed by means of an antibiotic, mitomycin-C, dormant bacteriophage fragments—phage "heads"—revealed themselves. "You can't really say any bacterial strain is free of phage," Dr. Pootjes remarked.

Her findings are additional evidence of the hide-and-seek relationship that certain viruses have with their host cells. When a virus invades a susceptible cell, it disassembles into its molecular components—including such large molecules as DNA. The viral DNA may take over the molecular machinery of the host cell, converting it into a factory for production of new virus particles. Alternately, the viral DNA may combine with the genetic material of the host, the chromosomal DNA, and remain dormant there as a "prophage." That action is presumed to occur in virally induced animal cancers, influencing the transformation of cells from a normal to a cancerous state. In a bacterial cell harboring phage DNA, the dormant prophage apparently does not injure the host, and is replicated and passed along to succeeding generations of bacteria.

Somewhat ironically, Dr. Pootjes, who showed that "cured" bacteria may be carriers of the virus in the prophage state, says she doesn't believe the virus is the elusive tumor-inducing principle. Instead, she suggests that the TIP might be a plasmid--a fragment of genetic material in the bacterial cell, which may be either bacterial or viral in origin, but is not part of the bacterial chromosome. "The current idea is that what goes from the bacterial cell to the plant cell is DNA. But they've been unable to come up with direct evidence of viral DNA, they've been unable to come up with direct evidence of bacterial chromosomal DNA, which would leave plasmid DNA," she told me. Another reason Dr. Pootjes gives for doubting the role of the virus is that no one has been able to induce tumors with viral DNA alone, in the absence of the tumefaciens bacteria.

A cautious supporter of the virus theory is Dr. Braun. "There is very little question that there is some self-replicating entity in these tumor cells that is transmissable. Certainly this is a viral-like entity," he told me. A plasmid, he said, "of course would be a virus in the broadest sense. I think any submicroscopic entity that is capable of self-replication would have to be placed in the broad category of viruses."

Rather than a DNA virus, the TIP could conceivably be an RNA virus. (RNA--ribonucleic acid--is a messenger molecule that copies DNA instructions and transmits them to appropriate cellular regions for chemical manufacture; in some viruses, RNA is the genetic core material.) Dr. Braun says that Pierre Manigault of France's Pasteur Institute claims to have induced tumors with an RNA extracted from the tumefaciens bacteria. "I haven't seen Manigault's evidence," he said, "but he's an experienced man who would at least be aware of the pitfalls. He claims these tumors that he induces with an RNA are transplantable, which is really the best criterion one has of the transformation."

THE CASE FOR BACTERIA

A number of researchers, however, look to the bacterium rather than the virus as the TIP. After all, crown gall is unique among cancers in being communicated by bacteria--perhaps it's also unique in requiring the activity of bacterial DNA. In theory, it's possible for bacterial DNA to migrate among a plant's cells, along the protoplasmic strands between the cells. Dr. Lippincott, although he claims to see tumor initiation as a "black box," reported evidence at the AIBS meeting supporting that model.

Wounding pinto bean leaves by rubbing them gently with carborundum powder, he then applied different strains of tumefaciens bacteria to various groups of plants. Chemical analysis of the resulting tumors showed that those plants infected by a particular bacterial strain synthesized the same "tumor growth factor." Lippincott calls these TGFs--octopine, nopaline, TGF-1 and TGF-2--phenotypes, or genetic markers.

These substances help promote tumor growth; more interesting, when bacteria-free tumor tissue is cultured from the various tumors, each "strain" of tumor cells continues to produce the same marker chemicals originally produced by particular bacterial strains. Because synthesis of these TGFs is controlled by DNA, and because the plant cells normally do not produce them, it's reasonable to suppose that bacterial DNA may be replicating away in the plant-cell chromosomes.

Dr. Robert Beardsley of Manhattan College says that this is insufficient evidence for a bacterial TIP; in his opinion, there may be an association of some sort between the bacterium and the bacteriophage, as in the case of diptheria: diptheria bacteria have to be infected with a specific phage in order to produce their disease toxin.

In experiments conducted with Dr. Judith Leff, Beardsley attempted to induce tumors on wounded plants with phage DNA extracts rather than with whole phages. "It would be surprising if plant cells possess the enzymes necessary to remove the capsids [the outer protein coat] from phage particles, thereby releasing the enclosed nucleic acid," he explains. There was some sporadic response, largely in the form of tumor-like tissue that reverted to a more normal condition or stopped growing after a while.

Dr. Beardsley says that it's possible the virus DNA "increases the frequency with which habituation occurs." Habituation is a phenomenon in which "normal tissue culture that has never seen a tumefaciens, growing in vitro, will become auxin-independent. [Auxin is a growth hormone normally required by plant cells.] And some of these are graftable--when you graft them to a normal host, they form tumors. But this phenomenon is non-reproducible, occurs at random, and you can't say one tissue culture out of a hundred is going to show it. Even though we could raise the level with which these tumor-like proliferations occurred, it did not occur consistently."

WHAT ABOUT THE WOUND?

Many researchers believe that conditions in the plant wound at the time of tumorigenesis indicate a cooperative or at least synergistic relationship between all three entities involved--the bacterium, the plant cell and the virus. Dr. Lippincott's research team, among others, has shown that the bacteria find "receptor sites" on cell walls adjacent to those ruptured in wounding. If the bacteriophage, or some part of it, is the TIP, it cannot enter the plant cell without the assistance of the bacterium, since phages are adapted only to entering bacterial cells. Tumor initiation begins in uninjured plant cells; although the bacteria do not enter the intact plant cells themselves, presumably they facilitate the entry of the TIP. Because washing the wound after inoculation prevents both tumor formation and normal wound healing, the wound sap presumably contains hormones that stimulate cell division in uninjured plant cells. It is during this regeneration of tissue that conversion to the neoplastic state begins: the chromosomes of the plant cells are opened up during cell division when DNA synthesis takes place, and are at that time apparently most vulnerable to the TIP.

The relationship between tumorigenesis and cell division was demonstrated in a classic series of experiments in the laboratory of Armin Braun in 1943; they also helped to show that transformation is not an all-or-nothing phenomenon--some tumor cells are "more cancerous" than others. Dr. Braun inoculated periwinkle stems with tumefaciens bacteria, then killed the bacteria by heat after varying intervals--24, 34, 60 and 72 hours. The longer the plant cells had been dividing in the wound-healing process before bacteria were eliminated, the faster-growing and more malignant were the tumors. Moreover, the "weakly induced" tumor tissue was less autonomous than fully transformed tumor cells, in that it required some of the plant cell nutrients to grow. This precisely parallels the cancer "progression" and the "hormone-dependent" cancers of animal research.

While some scientists argue over whether the TIP is bacterial DNA, viral DNA or both, others are coming up with evidence that neither hypothesis is true. Dr. Mary-Dell Chilton of the University of Washington told the Amherst meeting the results of an assay technique called DNA hybridization, in which DNA strands from separate organisms are recombined to see how well their nucleotide sequences match. It turns out, Dr. Chilton said, that neither <u>tumefaciens</u> DNA nor phage DNA meshes with tumor DNA. The work of Dr. Chilton and her associates is highly regarded among American crown gall researchers, so there is a tendency to feel that "the whole thing is up in the air," as Dr. Pootjes put it.

<u>REVERSING CANCER</u>

For Armin Braun, etiology of the crown gall cancer is less important than its potential as a model system for chemotherapy. He is pursuing lines of investigation leading from his 1950s demonstration that the cancerous state in crown gall tumor tissue is reversible. Using weakly inducing tobacco plant tumor tissue (a teratoma form which includes a variety of cell types), he transplanted bits of tissue descended from single tumor cells onto growing tobacco stems; after the grafting process was repeated through several generations, the plants flowered with normal seed. Other experiments demonstrating reversibility in animal cancers have been done in the last few years.

Demonstration of cancer reversibility, Dr. Braun told me, has caused an important shift in thinking about the neoplastic state. Irreversibility used to be dogma among cancer researchers, partly because it complemented the idea that carcinogenesis meant some damage to the cell's genetic material. There <u>are</u> some cancers in which that appears to hold true, he said, such as those brought about through exposure to radiation. But in general, the tumor problem is simply a problem of anomalous differentiation, according to Dr. Braun.

<u>GENETIC EXPRESSION</u>

"What is happening here is the same thing that happens in the normal course of differentiation in the embryo, with cells differentiating into kidney cells, liver cells, and so forth--except that you're getting <u>anomalous</u> differentiation. The evidence is very good that differentiation is based on differences in gene <u>expression</u>--that is, all the cells of a given organism are believed to contain the identical complement of genes, and it's simply a question of which of those genes are expressed as to what gives individual cells their characteristic properties." In the vernacular of molecular biology, some genes are repressed or "switched off," while others are derepressed or "switched on."

The tumor cells are also specialized--they are programmed to be parasitic on the total organism. "These things," said Dr. Braun, "are coded for cell division. Those that are fully transformed can produce, in optimal amounts, all the factors that are required for their continued rapid growth. Their essential purpose in life is to divide, and their whole metabolism is geared to this."

Dr. Braun has shown that crown gall cells produce their own auxins and a cell-division hormone, cytokinesin, which acts synergistically. Once the TIP has done its work, the transformed tumor cells' descendants perpetuate the same specialized metabolic abilities. It has been suggested that this inheritance of chemical traits does not require the inheritance of the original TIP. Dr. Braun disputes the proposal, citing an experiment by Dr. Ian A. Macpherson of the Imperial Cancer Research Foundation, London. Individual cells for virally induced hamster tumor tissue were cloned, and some of the descendants reverted to normal; the reversion was always accompanied by loss of the virus from the cells. "This shows clearly that the virus is necessary to maintain the pattern of synthesis that's characteristic of the cancerous state," said Dr. Braun. Presumably, the virus exerts some continued control over the genetic machinery of the cancerous hamster cell, keeping certain genes switched on or off.

A DEADLY PROGRAM

'Programming cells for death: is the idea behind Dr. Braun's chemotherapeutic approach to cancer research. Rather than being rigidly controlled by genetic instructions, many kinds of cells are now seen as "multipotential"--responding in various ways to hormonal, nutritional or other stimuli from the cellular environment. "The immediate challenge," says Dr. Braun, "is to find methods for completely replacing the pattern of synthesis concerned with autonomous growth by a pattern involved in differentiated function." In other words, to wrest control from the tumorigenic agent, whatever it may be.

'Ultimately we want to program tumor cells in such a way that we produce a terminal differentiation which will result in their death," Dr. Braun said. "This is not as farfetched as you might think, because it is very well known that cells are normally programmed for death in the development of the embryo. For example, the cells that join the fingers and toes during embryonic development of man and other mammals. These cells die out, and this permits the normal development of the embryo."

Terminal differentiation takes place naturally in certain cancer cells, he pointed out. In squamous cell carcinoma, a common skin cancer, interior tumor cells are inert and not carcinogenic, it has been proved that these cells are derived from the actively cancerous outer cells. A somewhat similar situation exists in normal plant tissue: certain multipotential cells die in an organized manner and become tracheids, tubes for conducting water.

Because the development of tracheids is obviously under hormonal control, Dr. Braun decided to investigate it. Working with pith parenchyma cells from romaine lettuce, he and his co-workers have learned how to program the development of the tracheary cells with hormones and related substances. A key chemical is cyclic-AMP, cyclic adenosine monophosphate, common to animal and plant cell regulatory processes, and the focus of new cancer research; a cyclic-AMP derivative promoted the terminal differentiation of the romaine pith cells. Dr. Braun is now trying to apply the same sort of control to crown gall tumor tissue as a model for cancer study.

Human cancer researchers familiar with the crown gall work speak of it with enthusiasm--even "unbounded admiration," in the case of Dr. Roy Albert of the Institute of Environmental Medicine, New York University Medical Center. He told me that Dr. Braun's research, specifically, is so important, it "may bear the same relationship to mammalian cancer as Mendel with his sweet peas bears to mammalian genetics. It may be a parallel situation in which the fundamental mechanisms are first worked out in plants and then picked up in mammalian systems, which are more complicated and difficult to study. It's certainly one of the most exciting aspects of cancer research, in my opinion."

Official recognition has come from the U.S. Senate's National Panel of Consultants on the Conquest of Cancer. Discussing gene expression--"anomalous differentiation" rather than genetic injury in the cell--as a cause of cancer, a panel report said: "Much more is known about how to 'derepress' certain genes in plant cells ...and also how to 'repress' them (as during tumor reversal). For these reasons, studies of plant tumor cells seem admirably suited for a first characterization, at the molecular level, of those mechanisms that may underlie not only their own, but all cancerous states."

*

CULTURED ORGANS

>Organ transplantation can have legal as well as medical problems.

The major problem facing routine organ transplants is the phenomenon known as rejection. To counteract rejection, doctors have resorted to immunosuppressive drugs, designed to suppress the body's tendency to attack the grafted organ. Furthermore, donor organs have been matched to specific recipients in hopes of combating rejection--a process similar to the way blood is typed before transfusion. On the whole, however, the problem of rejection has proved more formidable than originally thought. Proof of this is the near-perfect failure record of heart transplants.

Since the first transplants, scientists have been searching for alternate ways to combat rejection. Now it appears that Dr. William T. Summerlin, head of a transplantation biology effort at New York's Sloan-Kettering Institute, may have hit upon a solution.

Summerlin's startling find is that organs lose their ability to trigger rejection if they are preserved for a specific length of time in nutrient solutions ordinarily used to keep cells alive. He began "organ culture" four years ago, first keeping human skin vital in solution for up to eight weeks before grafting it onto hospitalized patients suffering from burns. The transplants were successful despite the lack of initial compatibility--a lack that would normally have led to rejection.

Something important was happening to the skin during the time spent in organ

---

Copyright © 1973 by Saturday Review Co. Used with permission.

culture. What that was no one now knows, but Summerlin has employed the technique to successfully graft pig skin onto mice, to transplant adrenal glands from one strain of mice to another, and even to transplant human corneas onto rabbits, all without the use of immunosuppressives.

Aside from skin transplants, the technique has not been directly applied to humans, but it offers some intriguing possibilities, among them restoring insulin balance to diabetics by pancreas transplants. Should the scientists discover what is actually happening in their culture dishes, the prospects are nothing short of fantastic.

TRIPLE TRANSPLANT DONOR SLAYING DILEMMA

PALO ALTO--Attorneys are girding for a legal battle over the definition of death in the case of a living heart transplanted from a young shooting victim whose brain showed no sign of life.

Stanford University Medical Center said the heart of 29-year-old Samuel Moore was beating steadily Thursday in the chest of a 52-year-old construction worker after a four-hour operation Wednesday.

Moore's transplanted kidneys also were keeping two women, 52 and 62, alive in San Francisco's Presbyterian Hospital as part of the unprecedented three-organ transplant.

The transplant operations have raised legal questions because Moore's kidneys and heart were removed before doctors disconnected the heart-lung machinery that kept him alive.

The Alameda County district attorney's office was expected to seek a murder indictment against A. D. Lyon, who initially was charged with assault with a deadly weapon in Moore's shooting.

But a decision had not been reached Thursday, and the case was complicated by Lyon's attorney, John F. Cruikshank, who countered the suggestion of a murder charge by asking:

"How could he (Moore) have been dead if his heart was still beating?"

"That's a very interesting question," said Dep. Dist. Atty. Joseph Salgado of Alameda County. "I'm afraid my expertise is limited and I don't know what the medical testimony in court will revolve around.

"That (the definition of death) is a trial problem, not a complaint problem

---

Used with permission of Associated Press.

at this time. In order to charge someone with murder there must be a death," Salgado said.

Alameda County's coroner and district attorney had agreed to risk the possibility of legal complications in order to save the life of the potential heart recipient.

Moore was shot in the head with a .22-caliber bullet three days ago but was kept alive by life-supporting machines. His wife and mother gave consent for the heart and kidney transplants after doctors at Highland Hospital in Oakland determined he would die of the bullet in his brain.

NEUROLOGICAL DEATH

Dr. Robert Burns, head surgical resident at Highland Hospital, said a neurosurgeon and a neurologist formally pronounced that Moore had suffered "neurological death" before the transplants were permitted.

Burns said no evidence of brain activity was visible in Moore's electroencephalogram readings over a 24-hour period. He said this conformed to criteria of death devised by Harvard medical school.

Stanford said they believe the heart transplant team headed by Dr. Norman Shumway performed a double first: the first heart transplant from a criminal shooting victim and the first involving helicopter transportation of the heart from one hospital to another.

The heart recipient, who asked to remain anonymous, had been bedridden with heart disease for months and was described as "super-critical" before the operation. He was reported in satisfactory condition Thursday.

HEART REMOVED

Shumway and Dr. Eugene Dong drove to Oakland from Stanford Wednesday morning, stopping to change a flat tire along the way. Two Highland doctors assisted them in removing Moore's heart, which was placed in a solution of cold brine that gives surgeons two hours at most to perform a transplant.

The heart was placed in the care of Dr. Alvin Hackle, an anesthesiologist-pediatrician familiar with helicopter emergency flights, and flown to Stanford where another member of the Shumway team had begun operating on the recipient.

Shumway returned to Stanford by car and arrived in time to assist Dr. Randall Griepp with the implant.

Meanwhile, Burns removed Moore's kidneys which were rushed across the Bay Bridge to Presbyterian Hospital.

MEDICAL ETHICS AND HUMAN SUBJECTS

For the sake of knowledge of life, scientists often kill, or permanently injure, living beings. Life scientists, as well as the rest of society, must constantly weigh the potential value of research against the pain it will bring.

The Tusgegee syphilis study went on for 40 years before public disclosure forced an investigation. Now the Department of Health, Education and Welfare has decided that the study was unethical, even when it started.

The sterilization of young girls in Alabama, while not a research project, is another case of questionable medical ethics. The Civil Rights Division of the Justice Department and the Senate health subcommittee are investigating charges that as many as 11 minors may have "undergone involuntary sterilization operations."

These highly publicized incidents "may give the erroneous impression that such procedures raising ethical questions are rare and involve only bizarre procedures ...This is not the case," say Robert M. Veatch and Sharmon Sollitto of the Institute of Society, Ethics and the Life Sciences in Hastings-on-Hudson, N.Y. In the June Report of the Hastings Center, they cite 11 studies they believe raise disturbing questions.

The studies were selected from a collection of 43 questionable experiments that have been published in responsible medical journals or professional proceedings since 1966. No names are mentioned but all the studies were performed in the United States or with funding from the United States.

The first three involved "grave risks to subjects." In one, nine normal female patients were given injections of epinephrine in an attempt to produce arrhythmia

Reprinted by permission of *Science News*, © 1973 by Science Service.

or abnormal heart beat. Those conducting the experiment admitted that such a procedure is hazardous, but they said, "informed consent cannot be obtained in a study of this type." In a second study, blood samples had to be taken from patients who had had both of their kidneys removed, some as recently as two weeks prior to the experiment. The ten subjects had to be transfused "in anticipation of blood loss due to repeated sampling." By the third day, "all subjects were clinically dehydrated." A third study involved giving LSD to 24 subjects in order to study long-range changes in personality, attitudes, values, interests and performance. No mention was made to the subjects of possible personality changes.

Four experiments involved risks to incompetent or incarcerated subjects. Nine children suffering from asthma were intentionally subjected to doses of antigens known to produce asthmatic attacks. In another experiment, 48 children suffering from blood diseases were subjected to dual-site bone marrow withdrawals. The study pointed out that such procedures "involve physical and psychological problems" for children.

In prisons or mental institutions, the quality of consent, even if it is obtained, is questionable. At a maximum security facility for the criminally insane, 90 patients were "used in an exploratory study to determine the effectiveness of succinylcholine as an agent in behavior modification." The drug causes temporary muscle paralysis, including inability to breathe.

Another experiment in operant conditioning was reported by an American psychiatrist working at a mental hospital in Vietnam. Chronic male patients (mostly schizophrenic) were offered freedom if they proved they could work. Of 130, only 10 volunteered to work. The rest were told they needed treatment and were given electroconvulsive shock. After a few treatments, most of the men decided to work. A similar experiment was then tried on 130 women. Even after each had received 20 treatments, only 15 were willing to work. Shock treatments were discontinued and food was withheld for periods of up to three days. The patients were eventually cured and went to work tending crops for the Green Berets.

In a placebo experiment, 91 of 130 children with bronchial asthma received injections of buffered saline instead of medication. This ineffective treatment lasted in some cases for 14 years. Neither the children nor their parents knew an experiment was going on. In an experimental birth-control program, 262 women had megestrol acetate capsules implanted in their forearms to test the long-term contraceptive effectiveness of the drug. This experiment produced 48 unwanted pregnancies. In a study involving legal and psychological risks, 332 patients in a voluntary psychiatric hospital were subjected to urine analysis for drug use. No consent was obtained and the patients did not know what the test was for. In a final experiment, 41,119 patients enrolled in a group health plan were given a test for pain tolerance as part of their regular checkup. They were subjected to as much pain as they could stand but did not know they were part of an experiment.

These examples, say Veatch and Sollitto, indicate the need for mechanisms for consent and review which give greater assurance that the rights and interests of subjects will be protected. The procedures now available, they say, are inadequate. In peer review, for instance, it is the peers of the researchers not the peers of the subjects who are asked to evaluate the ethical acceptability of the proposed research. Conclude the Hastings Center researchers: "The immediate establishment of a governmental committee to formulate rigorous procedures to ensure reasonable informed consent and review is the minimum that is called for."

This week, after testimony from the family involved in the Alabama sterilization case, the Senate health subcommittee began consideration of a bill that would strictly control medical experimentation. In a similar case, judges in Detroit ruled this week that psychosurgery may not be performed on prisoners or mental patients confined against their will.

*

TESTS ON ABORTED LIVE FETUSES BANNED

SACRAMENTO (UPI) Gov. Reagan Monday signed legislation banning medical experiments on live, aborted fetuses.

The bill by Assemblyman Mike Antonovich (R-Los Angeles) was designed to prohibit inhumane medical practices on fetuses which are kept alive in artificial wombs for experimentation.

The measure was vehemently opposed by Stanford University officials who said it would doom vital research which could save untold millions of future children.

During legislative hearings, James Babcock, a premedical student at UC Berkeley, testified that Stanford researchers in the past have subjected live fetuses to "grotesque torturing."

He cited one experiment where the rib cages of the fetuses were cut open to observe the heart beating under pressure. If the fetuses instead had been placed in an incubation chamber, they would have had a "fighting chance" to live, Babcock said.

Reprinted by permission of United Press International.

ACUPUNCTURE: THE WORLD'S OLDEST SYSTEM OF MEDICINE
John W. White

Modern medicine appears to have something to learn from the ancients.

"That a needle stuck into one's foot should improve the functioning of one's liver is obviously incredible...The only trouble...is that, as a matter of empirical fact, it does happen."

The writer of that startling observation was Aldous Huxley. The words come from his foreword to a 1962 book entitled <u>Acupuncture. The Ancient Chinese Art of Healing</u>, by a British medical doctor, Felix Mann. Today Dr. Mann is one of the foremost authorities on acupuncture, the ancient Chinese therapeutic and anesthetic treatment by needles that is making headlines in 1972. It was typical of Huxley's genius and vast learning to help introduce the West to the world's oldest system of medicine.

Huxley went on to describe how a skilled acupuncturist can cure, improve or arrest a wide range of afflictions by inserting needles into the body at various points and depths. A short listing would include migraine headache, ulcers, arthritis, high blood pressure, conjunctivitis, hay fever, acne, sciatica, convulsions, hepatitis, asthma, hemorrhoids, angina pectoris, lumbago, weak eyesight, tonsilitis, anemia, and insomnia. No surgery or drugs, mind you--just needles.

Huxley would be pleased to know of the controversy sparked by serious medical interest in needle puncture a decade later. With the reopening of China, a few Americans scientists have at last been able to observe acupuncture first hand and report on it in the medical literature. The December 6, 1971, issue of the prestigious <u>Journal of the American Medical Association</u> had a lead article which made

---

Copyright © July 1972 by *Psychic* Magazine, San Francisco. Used with permission of the publisher.

favorable pronouncements about acupuncture by Dr. E. Grey Dimond of the University of Missouri. Dr. Dimond had made a trip to China several months earlier, in company with eminent cardiologist Dr. Paul Dudley White and Dr. Samuel Rosen of Mount Sinai School of Medicine. A fourth medical man, Dr. Victor Sidel of Montefiore Hospital and Albert Einstein College of Medicine in New York, also observed medicine in China at that time. All three commented favorably about acupuncture to the press after their journeys.

Dr. Arthur Galston, a plant physiologist at Yale, and Dr. Ethan Signer, a molecular biologist at MIT, began the controversy. In May 1971, they made history by traveling to North Vietnam where they were the first American scientists to be granted interviews with Premier Pham Van Dong. They continued on to China for a similarly historic meeting with Premier Chou En-lai and exiled Cambodian prince Norodom Sihanouk. When they returned to America and met the press, they told reporters they had witnessed four major operations (hernia, ulcer, ovarian cyst, and thyroid tumor) performed in Peking using acupuncture as the only anesthetic agent.

That statement made national headlines. It also made some scientists uncomfortable and even angry. "Nonsense and rubbish," they said. "It's all in the mind. Needles instead of sugar pills. Hypnosis. Traditional Chinese stoicism. Trickery." It is no exaggeration to say that Drs. Galston and Signer were "needled" by some professional men in a most unprofessional way. They had challenged some basic beliefs of Western science.

Recently Dr. Galston told me, "Not all the effective points in acupuncture are consistent with the neural pathways. The Chinese know this and admit it. They say, in effect, 'We have an empirical body of data but we have no theory which will adequately explain the data in terms of what we now know about transmission pathways of pain and sensation. This may mean that there is some other kind of system which modulates sensation.' That, of course, is a shocking thing to tell a neurobiologist who believes that he understands all the parameters of his system. But when you've got a conflict between theory and empirical fact, it's the theory that has got to give way."

One of the empirical facts which better-informed observers point to is this: acupuncture works equally well on animals and infant children. Presumably both are inaccessible to hypnosis, autosuggestion, and placebo effect. That is acupuncture's challenge to the West: neurophysiological theory must explain it or else the theory itself must be revised.

What do we know about acupuncture so far? The name comes from Latin: <u>acus</u> --needle, <u>punctura</u>--puncture. Legend has it that acupuncture originated from the chance discovery by a soldier that arrows shot into one part of the body could cure illnesses and anesthetize other parts of the body. The beginnings of the system are lost in antiquity, but as early as 2600 B.C. the Chinese emperor Huang Ti gave it official status by telling his court physician to write the laws of acupuncture in a book for use by all future physicians. That book is <u>The Yellow Emperor's Classic of Internal Medicine</u>, the most ancient text available on traditional forms of Eastern medicine. Its only English translation was done by Dr. Ilza Veith in 1949, one of the foremost authorities on Oriental medicine in the Western world, and professor and vice chairman of the Department of the History of Health Sciences, University of California School of Medicine, San Francisco. Dr. Veith, who received the first doctorate in the history of medicine ever given in this country, says "I am not a crusader for acupuncture, but I am very anxious to see it studied in this country." And Dr. Veith has stated many times, as she did in her article in the <u>Journal of the American Medical Association</u> in 1962, that it behooves organized medicine to establish scientifically whether acupuncture is a valid medical procedure.

Interestingly, in the Emperor's time the needles were made of bone, antler, and porcelain. Later they were replaced by metal needles--gold, silver, and nowadays, stainless steel. The earliest needles made in neolithic times were of flint, and in The Yellow Emperor's Classic reference is made to nine kinds of fine needles, which were probably metal.

Since Huang Ti, acupuncture has been an ever-present and respected method of healing in China, Korea, and Japan--much as aspirin is in America. It wasn't until Jesuit missionaries returned from China in the eighteenth century that the West got its initial glimpse at this strange system of medicine. In 1928 the first accurate report of acupuncture was published in France, but not until the 1950s did the English-speaking world learn of it. Today there are more than a thousand doctors in western Europe, another thousand in Russia, and some three hundred in South America who are practicing acupuncture. Japan has more than 25,000 acupuncturists while China has about one million. Of that million, only 150,000 are physicians because in 1949 Mao Tse-tung decreed that every doctor be trained in acupuncture and all aspects of traditional medicine, even if he doesn't practice it.

The United States, however, has only a handful of medically licensed acupuncturists, although there are some traditionally trained practitioners in major cities who clandestinely pursue their livelihood in defiance of the laws which govern medical credentials. The lack of properly licensed acupuncturists may soon change, however, because in May 1971 the North American College of Acupuncture was established in Vancouver, Canada. It already has thirty students, including three MDs. The staff is mostly oriental.

Traditional Chinese medicine is intimately bound up with Chinese philosophy, Taosim and the yin-yang conception of dynamically opposing but harmonized energies in the universe. These energies wax and wane in a rhythmic fashion. Man is a microcosmic representation of the universe and, therefore, also has the same regular change in his body's energy. Because of this background, Chinese medicine is quite different from "objective" Western medicine. Acupuncture treatment aims at prevention, as well as cure. The skilled acupuncturist, using a complicated method of pulse diagnosis, determines the yin and yang condition of his patient and, if there is an imbalance, treats him for what he foresees will go wrong if not corrected. Should a disorder arise, he treats the person, not the illness. That is a simple statement, yet it expresses a universe of difference between East and West. How often do Westerners go to a doctor to prevent disease as compared to their visits to be "cured"?

Acupuncture theory says that the body has a vital energy called ch'i (sometimes spelled qi or t'chi) which circulates throughout it along a system of 12 bilateral lines or channels called meridians. Illness is due to a malfunction or imbalance of this energy system. Each meridian is associated with an internal organ such as the heart, lungs, stomach, etc. On the meridians are more than 800 points, each about 1/10 of an inch in diameter, which are the puncture points. These points are mapped out on charts of human and animal bodies for use by acupuncturists. (The classical number of points--365--has been expanded to nearly 900 in the recent past by acupuncturists who have experimented on themselves in response to Mao Tse-tung's exhortation during the Cultural Revolution to improve medicine.) By inserting fine needles into the appropriate points and varying the depth and speed of insertion, the acupuncturist claims he is affecting the energy flow, either stimulating it or dispersing it, in an attempt to restore equilibrium to the energy system. By restoring it to a condition of normal functioning, the patient is returned to health. Since the points of insertion avoid vital organs, the needles do no damage to the body although they may cause a little soreness.

According to the millenia-old theory of acupuncture, there is a proper time of day and night for treating each organ, depending on its periodic increase and decrease in energy. In the last few years, Western science has confirmed the concept of "body time," of which the most important for medicine is the circadian (about a day) rhythm. Pulmonary afflictions, for example, are largely confined to the early morning hours while cardiac disease is more common later in the day. Surgeons are aware of changing susceptibility in patients to drugs and anesthetics, along with regular shifts in bacterial resistance, temperature, sensory keenness, muscle tone, blood sugar, and dozens of other factors. Ideally, they all should be taken into account when planning surgery or the administration of medicine.

An alternative to acupuncture treatment is moxibustion or moxa treatment. In moxibustion, small cones of powdered Artemisia vulgaris leaves (commonly called wormwood or mugwort) are placed on the acupuncture points and ignited. When they have burned down far enough to form a small blister on the skin, the cones are removed. As with acupuncture, this procedure is supposed to restore the circulation of ch'i to normal and thereby give the same healthful effects as using needles. Massaging the acupuncture points is still another alternative, but it is less widely used than needling or moxibustion. The Chinese practice called "do-in" uses body massage or acupuncture points for improving muscle tone.

As mysterious as acupuncture is the related pulse diagnosis method. By feeling the pulse at three spots along the radial artery of each wrist, the acupuncturist claims to determine the functioning of different internal organs and various physiological processes. Each spot has a deep and superficial reading, making six on each wrist for a total of twelve. By the fullness or feel of the pulse, the state of yin and yang energy can be determined--for example, whether an organ has a deficit, or surplus of energy. Deep in the left wrist, science writer Walter Rose wrote in Medical Opinion in November 1971, the acupuncturist feels the pulses of the kidneys, liver, and heart. At these same spots, a superficial reading tells him about the bladder, the gall bladder, and small intestine. Acupuncturists say pulse diagnosis can be learned only by feeling pulses in the presence of someone who is already initiated into it. They also claim it is possible to differentiate the feel of the pulse before and after inserting needles.

Some of the ailments which acupuncture can treat were listed at the beginning of the article. There are many more. In Acupuncture: Cure of Many Diseases (a simplified version of his earlier book), Dr. Mann takes two pages to list illnesses which he has cured or helped, but adds that his catalog is much smaller than those in some other books. "This list," he writes, "is only a representative selection, which mentions relatively more of the easier-to-cure diseases than the very difficult ones."

Recent newspaper headlines show that the list may grow even larger as the Chinese continue to research and apply acupuncture. "Acupuncture Helps the Deaf in China" reported The New York Times last July. The story states that the use of acupuncture to treat deafness was discovered in 1968 by an army medical team which located the points that affect hearing by repeated experiments on themselves. The doctors say they have been successful in 90 percent of those cases resulting from a childhood disease. They offer as proof eleven children who they say were deaf and mute prior to 1969, but now are completely cured and have been taught to speak.

An even more startling advance in acupuncture was reported by The Los Angeles Times last year: "Chinese Treat Mental Ills With Acupuncture." This new therapy combines acupuncture with traditional herbal medicine, modern drugs, and doctor-patient discussions. The Chinese claim it cured 79.2 percent of the inmates including some who had been deranged for more than 20 years at a mental hospital in

Hunan province.

Acupuncture also is used on the Chinese as a means of anesthesia both for surgery and for simple pain relief from maladies such as stomach ache. Classically, the needles were superficially placed in the skin and left for 10 to 30 minutes. But with the coming of the Cultural Revolution, acupuncturists developed a new anesthetic technique. Now the needles are sometimes placed deeper--up to two inches. In addition, they are constantly manipulated in a rapid up-and-down motion (about 120 times a minute) through a distance of about 1/2 inch while being twirled between the thumb and fingers. Occasionally needles are dispensed with altogether in favor of direct, heavy pressure from a finger on the acupuncture point.

The most dramatic departure from tradition took place within the decade when electroacupuncture was developed, primarily by a woman, Chi Lien. In this procedure, Dr. Galston reported, the implanted needles are attached to delicate wires leading to an electrical junction box whose power source is a five-volt battery. The patient received 0.5 milliampere current for 20 minutes, remaining awake throughout while the needles are manipulated either by hand or by machine. At the end of that time, the patient is completely anesthetized in the area to be incised for as long as nine hours.

Proof of the efficacy of acupuncture anesthesia was offered by Audrey Topping, wife of <u>The New York Times</u> assistant managing editor Seymour Topping. The Toppings went to China in 1971 at the time Drs. Galston and Signer were there. Mrs. Topping was an eyewitness to open heart surgery performed on a woman whose only anesthetic was acupuncture. During the operation the surgeon actually held her heart in his hands for all to see. The woman, who was calmly sipping orange juice through a straw, apparently felt no pain and smiled at the observers. "We almost fainted," said Mrs. Topping. Some people did faint when Prof. Galston showed a color film, recently obtained from a Chinese source, of major surgery being performed with acupuncture anesthesia alone. One patient had a brain tumor removed. Another was operated on for a spinal defect. In both cases the patients were fully awake and apparently unable to feel any pain. Needles were used, but in the case of a dental extraction, only direct finger pressure on an acupuncture point in each cheek was used.

There are advantages to acupuncture anesthesia which are not always present in Western methods. Dr. Dimond noted that it is "absolutely safe," there is no interruption with the patient's level of body fluids, there is no postoperative nausea or vomiting, and the patient can still eat and drink. Dr. Galston points to an even more significant aspect: the mortality rate associated with general anesthesia. "It is very easy to put a patient to sleep," he said, "but not always so easy to wake him up. Thus, if a convenient and safe technique for local anesthesia were available, it might become very useful here."

The Russians are also actively using acupuncture. Dr. Stanley Krippner, a psychologist at Maimonides Medical Center in Brooklyn, New York, and Richard Davidson, a research assistant at Maimonides, visited the Soviet Union last year to observe Russian acupuncture research and other aspects of their science. "Their whole consideration of acupuncture seemed very straightforward and reasonable to me," Krippner said. "I was quite impressed with the way they were going about it."

According to Dr. Krippner, the emphasis in Russian acupuncture is on application. Several Moscow hospitals have acupuncture units. They use it for anesthesia and--in combination with other techniques--for treatment of various internal disorders such as liver, kidney, and stomach problems.

Sexual disorders are being treated through acupuncture by Dr. G. S. Vassilchenko, a Moscow physiologist and sexologist. His successful applications including bedwetting, impotence, and frigidity. "However," said Dr. Krippner, "he prefers to use various ointments which he massages into the acupuncture points, rather than needles. He mentioned one preparation called chloratile as being especially effective for bedwetters if it is applied at an acupuncture point toward the base of the spine." Dr. Krippner later checked this point on an acupuncture chart and found that it coincides with a point traditionally associated with the bladder.

This use of massage and ointments highlights some major differences between the Russian and Chinese approaches to acupuncture. The Russians rarely use needles but rely instead on electrical stimulation, massage, salves, ointments, and are now experimenting with laser beams. "We were told that epileptic fits have been stopped by directing a laser beam toward the acupuncture point on the septum at the base of the nose," Dr. Krippner remarked.

The Russians have found that there are individual differences among people which affect the placement of acupuncture points. These differences would seem to be genetic since Caucasians have some points located at slightly different places than Orientals. More and more acupuncture points are being discovered by both the Russians and Chinese as acupuncture technology becomes more sophisticated.

A device invented by the Soviet physicist Dr. Victor Adamenko actually measures the energy flow between acupuncture points. This transistorized instrument, called a tobioscope, is about the size of a pencil flashlight and lights up when passed near the skin over acupuncture points. American scientists such as Dr. William A. Tiller at Stanford University are now experimenting with similar devices.

The tobioscope is often used in conjunction with a high-frequency type of bio-energy photography (called Kirlian after its inventors, Semyon and Valentina Kirlian); a modified version of Kirlian photography is being done at UCLA by Dr. Thelma Moss and Ken Johnson (see their article in this issue).

By using the tobioscope to locate the acupuncture points and then photographing with the Kirlian technique, the Russians say that they are able to determine the health of an individual by the color of the points. "And their parapsychologists claim," notes Dr. Krippner, "that in some cases healers have been able to concentrate on acupuncture points and assist the individual back to health."

What explanation of these observations can be made? Last December Dr. Mann came from England to the United States under government sponsorship for lectures at the National Institute of Health in Washington, D.C. and State University of New York's Downstate Medical School in Brooklyn, New York. While at the medical school, The New York Times reported, Dr. Mann told an audience of 300 doctors, medical students, nurses and others: "Doctors are a funny breed. If they don't know how a thing works, they come to believe that it doesn't work. Now, I can't tell you how acupuncture works. All I know is that I practice it on my patients and an astonishing number get better."

The ailments he has successfully treated are not among the most serious ones, Dr. Mann noted. Excellent results are possible with diseases which are functionally or physiologically reversible. It is not of any value for diseases which are irreversible or for which there is a definite anatomical or physiological lesion such as a broken bone. He doesn't treat cancer, for example, although in some cases the pain of cancer has been relieved.

Dr. Mann was invited to the medical school by Dr. John W. C. Fox, assistant professor of anesthesiology there, who made a videotape of the presentation for use by medical people. I asked Dr. Fox if he could explain acupuncture.

"I have no theory of my own," he said. "Western physicians are not at all satisfied with the classical Chinese theory and want to explain acupuncture in terms that are readily understood or would fit in with our ideas based on cellular pathology and molecular biology. The hangup is that some of the organ systems of classical acupuncture use the same names as we do for anatomy but some of them are not known by our conventional anatomy. For example, there is one called 'the triple warmer.' A direct translation from the Chinese is 'three burning spaces.' This organ has something to do with metabolism or nutrition. It might be considered part of the stomach function. The triple warmer is very hard to understand or explain, yet it has a pulse position and can be diagnosed and treated."

Equally puzzling is the meridian system. Having no apparent correspondence with Western physiology, it nevertheless is of paramount importance to the theory of acupuncture. Is there some physical foundation to the concept?

Perhaps so. In 1962, Kim Bong Han, a Korean professor of physiology announced that he had discovered that the human body has a system of cell-groups just below the skin which are connected by thin tubular cells. This system has no connection with the blood, lymph, or nervous systems. The cell-groups and connecting structures correspond amazingly with the classical Chinese and Korean descriptions of acupuncture points and meridians. Electrical measurements made at the cell-group sites indicated, in Prof. Han's judgment, that they are indeed acupuncture points. (See an appendix to Dr. Mann's first book on acupuncture and an article by Prof. Han entitled "On the Kyungrak System," which appeared in Journal of the DPRK Academy of Medical Sciences on November 30, 1963, in Pyongyang, Korea.) However, according to Dr. Dimond in his JAMA report, a team of Chinese went to Pyongyang but were unable to confirm Prof. Han's findings.

In view of such large obstacles to understanding, most Western scientists find it more comfortable to ignore the matter—and have until now. But not entirely. In Europe, acupuncturists have been working diligently for many years to improve the theory and practice of their profession. There is an International Society of Acupuncture with headquarters at Aubersville, France. Since 1955, an annual Congress on Acupuncture has been held—in 1971 at Baden-Baden, Germany—where practitioners and researchers come together to exchange and evaluate new developments. The French, Dr. Fox says, displayed a fiber optic device with which they are beginning to be able to objectively detect differences in the pulse pressure wave—and thereby perhaps confirm pulse diagnosis. They also have designed a computer-like instrument into which a person can feed the diagnostic findings by classical pulse diagnosis. The computer then translates them into Western terms for orienting a classical acupuncturist to what organ systems he should treat. Another new development is applying high-frequency sound to acupuncture points—so called sonopuncture.

Of importance on the theoretical side is an article entitled "Acupuncture Verified By Electronics" in the November 1971 issue of the French journal Science et Vie. The article proposes that small electrical potentials on the surface of the body, measureable in millivolts at small specific locations, may be what the Chinese call the ch'i energy at acupuncture points. At those points electrical resistance is inexplicable and sharply lower. Hence they are sites of greater conductivity.

Another electronic theory of acupuncture is offered by a French physician, Dr. George Cantoni. Dr. Cantoni has found that normal people in good health have a d.c. electrical potential difference of 30 to 40 millivolts between the head and fingertips. If one's health is less than good, there will be a decrease in this voltage. This electrical balance or imbalance is his idea of what the Chinese mean by "the circulation of energy."

Perhaps the most promising Western explanation of acupuncture anesthesia has been offered by Dr. Ronald Melzack, a neurophysiologist at McGill University in

Montreal. It is based on his concept of the "gate control" of pain by inhibitory mechanisms in the spinal cord which allow or block the transmission of impulses which, when they reach the brain, are interpreted as pain.

In this model of pain, according to Dr. Fox, there is a possibility that certain peripheral stimuli such as a needle prick or moxibustion could cause the abolition of pain by altering the transmission of stimuli that give rise to pain when they are perceived by the brain. Supporting this, says Dr. Fox, is Dr. Felix Mann's very exhaustive review of the literature. Through it he has found experiments showing that small stimuli applied to the surface of the body of experimental animals and humans can produce marked effects on viscera. "This hypothesis requires no reference to yin and yang or to a mysterious energy circulation," Dr. Fox said. "It is on this type of explanation that I think acupuncture will probably become part of general medicine and anesthesia in the West."

This explanation would not completely satisfy Russian scientists, however. Their theory of "bioplasma" is held to explain a number of unusual events, including psychic phenomena. Bioplasma, they say, is a fourth state of matter which constantly interacts with other states of matter. The phrase "fourth state of matter" is linguistically compatible with the scientific materialism officially embraced by the Soviet Union. In effect, though, bioplasma is a new form of energy--biological energy--different from electrical, magnetic, or mechanical energy. Russians refer to the bioplasma theory rather than the Chinese yin-yang conception to explain the operation of acupuncture. The first English-language report of their acupuncture research was published this Spring in the English Journal of Paraphysics by Dr. Victor Adamenko, the Soviet physicist who invented the tobioscope.

The Chinese are aware of this research, according to Dr. Galston, but they think the Russian approach is an incorrect one. "They said they didn't think that was the way to go at it," he told me. "Maybe this is just because Peking and Moscow are not sending on the same frequency these days."

So ideological differences dominate the supposedly impartial world of science as it searches for an explanation of what is presently an inexplicable phenomenon. But even those differences cannot long retard a growing movement among scientists around the world to re-examine the nature of man and his bioenergetic systems.

Acupuncture's role in this was assessed by Dr. Galston in the Winter 1972 issue of The Yale Review: "For if the results (of acupuncture) cannot be explained solely on the basis of the nervous system, then we may have to invoke other systems or modalities which can control the sensation of pain. This could result in new insights into the operation of the human body, or it might end up with a relatively trivial explanation. But since the Chinese seem very happy to blend Western medicine with traditional Chinese practices, should we be less willing to learn from the wisdom of the East?"

PART VII
BEHAVIOR

FLO'S SEX LIFE
Jane Goodall

Jane Goodall has taken an intimate approach to the study of the behavior of our fellow animals.

Sex appeal, that strange mystery, is a phenomenon as inexplicable and as obvious among chimpanzees as human beings. Old Flo, bulbous-nosed and ragged-eared, incredibly ugly by human standards, undoubtedly has more than her fair share of it. At one time I thought it was simply because she was an old and therefore experienced female that the males got so excited when she became sexually attractive. Now I know better, for there are some old females who are almost ignored at such times, and some young ones who are courted as frenziedly as Flo.

When a female chimpanzee comes into heat—or into estrus, as a scientist would say—the sex skin of her genital area becomes swollen. Such swellings vary somewhat in size; some females develop a pale pink protuberance that is fully as large as a three-pint bowl, whereas others have smaller protuberances. A swelling usually persists for about ten days before becoming flabby and wrinkled and then shrinking away to nothing again. Normally it occurs at a point midway between menstrual periods, which in the female chimp occur about every thirty-five days. It is during her period of swelling that a female—whom I refer to frivolously as a "pink lady"—is courted and mated by the males.

By the time I returned to the Gombe Stream after my second sojourn at Cambridge—together with Hugo, who had, after all persuaded the National Geographic Society that we needed more film on chimpanzee behavior—Flo and two of her three offspring had become fairly regular visitors to camp. Flo's daughter Fifi was about

---

From *In the Shadow of Man* by Jane van Lawick Goodall. Reprinted by permission of the publisher, Houghton Mifflin Company.

three and a half years old then. She still suckled from her mother for a few minutes every two or three hours, and jumped occasionally onto Flo's back, particularly if she was nervous or startled. I knew, too, that she still shared her mother's nest at night. Figan, Fifi's senior by some four years, had just attained puberty. Some young males of this age are fairly independent of their mothers, but Figan spent most of his days traveling about with Flo and Fifi. Faben, Flo's eldest known son, was seldom seen with his family that year; he was an adolescent of about eleven years at the time.

When Hugo and I first returned to the Gombe Stream, Flo, Fifi, and Figan were still rather apprehensive when they came to camp. They spent most of the time lurking in the thick bushes around the clearing, only emerging to seize the bananas we put out for them. Gradually, however, they relaxed, particularly when David or Goliath was with them, and then they spent longer and longer out in the open. I still spent most of my time climbing the mountains, but when most of the chimpanzees moved far to the north or to the south, then I often stayed in camp, hoping that Flo would come.

Even in those days Flo looked very old. She appeared frail, with but little flesh on her bones, and thinning hair that was brown rather than black. When she yawned we saw that her teeth were worn right down to the gums. We soon found out that her character by no means matched her appearance: she was aggressive, tough as nails, and easily the most dominant of all the females at that time.

Flo's personality will become more vivid if I contrast it with that of another old female, Olly, who also began to visit camp at that time. Olly, with her long face and loose wobbly lips so sadly reminiscent of William, was remarkably different. Flo for the most part was relaxed in her relations with the adult males; often I saw her grooming in a close group with two or three males out in the forests, and in camp she showed no hesitation in joining David or Goliath to beg for a share of cardboard or bananas. Olly, on the other hand, was tense and nervous in her relations with others of her kind. She was particularly apprehensive when in close proximity to adult males, and her hoarse, frenzied pant-grunts rose to near hysteria if high-ranking Goliath approached her. She had a large pendulous swelling in the front of her neck which looked exactly like a goiter. It may have been one, since they are not uncommon among African women in the area; and if so it might account for much of her nervous behavior.

Olly tended to avoid large groups of chimps and often wandered about with only her two-year-old daughter Gilka for company. Sometimes she was accompanied by her eight-year-old son Evered; in fact it was he who first led his mother to our camp after coming several times himself with David and Goliath. Often, too, Olly and Flo traveled about together in the forests, and all four children were playmates of long standing. For the most part the relationship between Flo and Olly was peaceful enough, but if there was a single banana lying on the ground between them the relative social status of each was made very clear: Flo had only to put a few of her moth-eaten hairs on end for Olly to retreat, pant-grunting and grinning in submission.

Once a game between Flo's son Figan and Olly's son Evered turned into a serious squabble--as happens only too often when young males play together--and Flo rushed over at once, her hair on end, in response to Figan's loud screams. She flew at Evered and rolled him over and over until he managed to escape and run off screaming loudly. Olly hurried up too, uttering threatening barks and looking extremely agitated, but she did not dare join in, and so contented herself when all was over with approaching and, as though to appease the dominant female, laying a hand gently on Flo's back.

Flo was a far more easygoing and tolerant mother than Olly. When Fifi begged for food, whimpering and holding out her hand, Flo nearly always let the child take a banana. Sometimes she actually held one out to her. True, there were a few occasions when Fifi tried to take her mother's last banana and Flo objected; then mother and daughter rolled over on the ground as they fought for the fruit, screaming and pulling at each other's hair. Such incidents were rare, though, Gilka would never have dared to try such a thing. We rarely saw her even begging for Olly's bananas, and when she did she was usually ignored. She seldom got more than a taste of the skins until she got tame enough to come right up to us so that we could smuggle her a whole banana. And even then Olly often rushed up and wrenched the fruit away from her daughter.

Despite Flo's normally relaxed relation with the adult males, she did not usually compete for bananas with David or Goliath; rather, she waited until they had taken an armful before she ventured to attempt to get some for herself. And so, one morning in July 1963, Hugo and I were surprised to see Flo rushing up to join in as Goliath and David approached a pile of bananas. Then we saw that she was flaunting a large sexual swelling.

After seizing a heap of bananas but before taking a single bite, Goliath stood upright with all his hair on end, stared at Flo, and swaggered from foot to foot. As Flo approached, clutching some bananas herself, Goliath raised one arm in the air and made a sweeping gesture through the air with his banana-filled hand. Flo crouched to the ground, presenting Goliath with her pink posterior, and he mated her in the typical nonchalant manner of the chimpanzee, squatting in an upright position, one fruit-laden hand laid lightly on Flo's back and the other resting on the ground beside him.

Chimpanzees have the briefest possible intercourse—normally the male remains mounted for only ten to fifteen seconds. Nevertheless, before Goliath had done with Flo, Fifi was there. Racing up she hurled herself against Goliath, shoving at his head with both her hands, trying to push him off her mother. I expected Goliath to threaten the child, to hit at her, or at least brush her aside. Instead he merely turned his head away and appeared to try to ignore Fifi altogether. As Flo moved away, Fifi followed, one hand laid over her mother's swelling, looking back over her shoulder at Goliath, who sat eating his bananas. For a moment Fifi stayed close to Flo, and then she went off to scrounge a banana.

A few minutes later David Graybeard approached Flo with his hair bristling. Sitting on the ground, he shook a little twig, staring at her as he did so. Flo instantly ran toward him, turned, and crouched to the ground; again Fifi raced up and pushed and shoved at David as hard as she could. And David also tolerated her interference.

After this the group settled down. David groomed Flo and then, since it was a hot morning, lay down to snooze. Goliath followed suit, and everything was peaceful. Not long afterward we saw Evered move slightly away from the group, looking over his shoulder at Flo, who was watching him. Then he squatted down with his shoulders hunched and his arms held slightly out from his body. This was the typical adolescent male courtship posture, and Flo responded at once, approaching and presenting as she had for the adult males. Goliath and David both looked up but otherwise ignored the incident. Fifi, as before, ran up and pushed at Evered, and he too ignored her.

The next day Flo arrived very early in the morning. Her suitors of the previous day were with her; once again they courted and mated with her before eating their bananas; once again Fifi rushed up each time and pushed at them. And then, from the corner of his eye, Hugo saw another black shape in the bushes. As we peered we saw another, and another, and another. Quickly we withdrew into the tent and looked through binoculars into the vegetation. Almost immediately I recognized old Mr. McGregor. Then Mike and J.B. Also there were Huxley, Leakey, Hugh, Rodolf, Humphrey—just about all the adult males I knew. And there were some adolescents, females and youngsters in the group as well.

We remained inside the tent. Soon Flo moved up into the bushes and there was mated by every male in turn. Each time, Fifi got there and tried to push her suitor away. Once she succeeded: she jumped right onto Flo's back while Mr. McGregor was mating her mother, and pushed so hard that he lost his balance and tumbled back down the slope.

For the next week Flo was followed everywhere by her large male retinue. It was impossible for her to sit up or lie down without several pairs of eyes instantly swiveling in her direction, and if she got up to move on, the males were on their feet in no time. Every time there was any sort of excitement in the group—when they arrived at a food source, when they left their nests in the morning, when other chimps joined them—then, one after the other, all the adult males mated Flo. We saw no fighting over this very popular female; each male simply took his turn. Only once, when David Graybeard was mating Flo, did one of the other males show signs of impatience: irascible J.B. started to leap up and down as he swayed a large low branch so that the end beat down on David's head. But David merely pressed himself close to Flo and closed his eyes—and J.B. did not attack him.

The few adolescent males of the group, however, did not stand a chance. Normally these youngsters are able to take their turn, provided they wait until a flurry of sexual excitement has died down and the adult males are calm and satiated. Then, if an adolescent hunches his shoulders or shakes a branch at the female from a discreet distance she will usually move toward him in response. The older males, although they may glance at the mating pair, seldom object to these expressions of youthful passion. Flo, however, was an exceptionally popular female. We did sometimes see an adolescent hunching at Flo from behind some tree, but, though Flo herself usually got up and wandered toward him, several of the adult males instantly followed. It was as though they feared that Flo might try to escape. The proximity of his elders effectively nipped the amorous intentions of the adolescent at the start and he would retreat hastily to gaze at Flo from a safer distance.

Once we saw Evered sitting some distance from Flo and every so often glancing from her toward one or other of the adult males. It looked as if he was torn between desire and caution. He took a few tentative steps toward Flo and suddenly went off at a tangent, flicking at pieces of rock, throwing little handfuls of grass in the air, kicking at stones with his feet. Then he sat down and for the next ten minutes occasionally shook a branch as though to relieve his feelings.

On the eighth day of her swelling Flo arrived in camp with a torn and bleeding bottom. The injury must have just occurred. Within another couple of hours her swelling had gone. She looked somewhat tattered and exhausted by then and we were relieved for her sake that everything was over. At least we thought it was over, for normally a swelling only lasts about ten days. But five days later, to our utter astonishment Flo was fully pink again. She arrived, as before, with a large following of attendant males. This time her swelling lasted for three consecutive weeks, during which time the ardor of her suitors did not appear to abate in any way.

During this second period of pinkness we noticed that a strange sort of relationship had grown up between Flo and one of her suitors—a relationship of a type we have never seen since. The male was Rodolf (his real name is Hugo, but two named Hugo in one book are too confusing, and so I have given him Hugo's second name). Rodolf, in those days, was a high-ranking and enormously big and powerful chimpanzee, and he became Flo's faithful escort. He walked everywhere just beside or behind her, he stopped when she stopped, he slept in the nest closest to hers. And it was to Rodolf that Flo often hurried when she was hurt or frightened during those weeks, and he would lay his hand reassuringly on her or sometimes put one arm around her. Yet he did not protest in any way when other males mated with Flo.

During the final weeks Fifi became increasingly wary of the males; perhaps finally she had been threatened, or even attacked, by one of Flo's suitors. Whatever the reason, her gay assurance in life was gone for the time being. She stayed farther and farther away from the center of a group during any sort of excitement and she completely stopped interfering with her mother's sex life. She didn't even dare join a group that was loading up with bananas in camp--she who a couple of weeks earlier had actually taken fruits from the hands of the males that she tried to push from Flo.

The fact that Flo's milk had quite obviously, dried up with the appearance of her sexual swelling may have affected Fifi's behavior too. A quick suck of milk, more than anything else, seems to calm an infant chimpanzee. If one of the males threatened Fifi, she could still run to Flo, still be embraced by her mother--but where was the comfort of a sudden flow of warm milk? Whenever the group was resting and peaceful during those hectic three weeks, Fifi went close to Flo and either groomed her or simply sat beside her with one hand on her mother's body. And when the group was on the move, Fifi, instead of running jauntily along ahead of Flo or bouncing along behind, took to riding again as though she were an overgrown baby. Not only did she frequently perch ridiculously on her mother's back, but she sometimes actually clung on in the ventral position under Flo's tummy with her back bumping along the ground.

One day Flo, with Fifi on her back, came into camp alone. Her fabulous swelling had gone, shriveled into a limp flap of wrinkled skin. She looked worn out, faded, incredibly tattered after her strenuous five weeks. There were two extra pieces torn from her ears and a variety of cuts and scrapes all over her body. That day she just lay around camp for several hours, looking utterly exhausted. She and Fifi left as they had arrived, on their own.

The following day a group of males was in camp when Flo appeared plodding along the path. The moment they spotted her they leaped up, their hair on end, and charged off to meet her. Flo, with a hoarse scream, rushed up the nearest palm tree. The males ran on, with David Graybeard for once in the lead. They stopped under her tree and gazed up for a moment before David, slowly and deliberately, began to climb the trunk. What took place we shall never know, for leaves hid the two from sight. A few moments later David reappeared and climbed slowly down the trunk again. He walked past the other five males and plodded back to camp. A moment later Flo reappeared and moved down toward the waiting group. She hesitated for a second before finally stepping to the ground. Then, crouching slightly, she turned and presented her shriveled posterior to her former suitors. Goliath carefully inspected the flabby skin, poking around for a moment and intently sniffing the end of his finger. Then he followed David back to camp, and it was Leakey's turn to inspect Flo. After Leakey came Mike and Rodolf, and finally old Mr. McGregor. Then they all trailed back to their interrupted meal, leaving Flo standing on the path staring after them. Who can tell what she was thinking?

After that Rodolf again attached himself to the old female, and for the next fortnight continued to travel around with her and her family. Figan, who had stayed away from his mother during most of her pinkness, was back with her also. One day, about a week after the last vestiges of Flo's swelling had subsided, Rodolf, who had been grooming her, suddenly pushed Flo roughly to her feet and feverishly inspected her bottom, repeatedly sniffing his finger, an eager glint in his eye.

Obviously her hormonal secretions gave no indication of imminent pinkness, and after a few moments he permitted Flo to sit down again and continued to groom her. We saw him behave thus on three different occasions. But Rodolf would have to wait close to five years before Flo again went pink.

THE BIOLOGICAL DEPTHS OF LONELINESS
Robert J. Trotter

Seeing the misery of these creatures, one wonders if the new knowledge is worth their pain.

Hundreds of poets have used thousands of words to describe the bitter emotions that often accompany loneliness. Whether it be the alone-in-a-crowd feeling, separation from kith and kin or actual physical isolation, almost every individual has experienced the forlornness that comes from feeling truly separated from the world. Solitude is often relaxing and peaceful, but too much of it can produce boredom and lethargy. In extreme cases, isolation can cause mental stress that leads to anxiety, depression and even psychosis.

The reason isolation can have such profound effects on psychological disposition is basic. Any change in the environment works a change in the organism.

Isolation usually implies a decrease in environmental or sensory stimulation. This decrease produces subtle biochemical fluctuations in the brain and nervous system and, in turn, changes in mental processes.

The outward manifestations of these biochemical changes have been recorded many times in animals. Hoyle Leigh of Yale University School of Medicine and Myron A. Hofer of Montefiore Hospital in New York reported behavioral and physiological changes in young rats after isolation from their littermates (SN: 4/29/72, p. 281). Harry Harlow of the Regional Primate Research Center at the University of Wisconsin studied isolated monkeys (SN: 8/1/70, p. 100). He found that six months of social isolation results in seemingly permanent abnormal social, sexual and maternal behaviors. High levels of aggression were particularly apparent.

Reprinted by permission of *Science News*, © 1973 by Science Service.

William T. McKinney Jr., and his co-workers, also at the Wisconsin Primate Center, noted that isolated monkeys show severe persistent psychopathological behaviors similar to those of autistic children (SN: 5/13/72, p. 311).

McKinney has been able to achieve partial rehabilitation of these animals by two methods. In one experiment, attempts were made to reestablish biochemical balance by setting up a therapeutic social environment. The isolates, who fought with peers, were put with younger animals and began to show signs of returning to normalcy. In another effort, the researchers attempted to work directly on the brain by administering drugs. Chloropromazine was the most effective. The biochemical substances implicated in these experiments are the catecholamines such as epinephrine and norepinephrine and others that are known to exert an important influence on nervous system activity. But to reestablish normal chemical balance, more must be known about which catecholamines and other substances in the nervous system are involved and whether isolation causes increases or decreases in secretion of them.

Answers to these questions could establish a biochemical explanation for psychotic behavior. McKinney has recently sent brain specimens from disturbed monkeys for biochemical analysis. The animals involved had been separated from their mothers. They showed signs of hyperactivity (excessive vocalizing) followed by a period of depression. Preliminary data from analysis of these animals' brains show that major biochemical changes do accompany separation. "There is a lot of evidence," says McKinney, "to reflect changes that would produce nervous system activation both peripherally and centrally." McKinney is not ready to say exactly what is going on but he hopes to have a more definitive analysis ready for publication in the near future. He and his co-workers will then proceed with brain-chemical analysis of monkeys who have been completely isolated rather than just separated from their mothers.

Leigh and Hofer's experiments with separated rats leads them in the same direction. The changes in physiology (increased heart rate), says Leigh, "suggest that the autonomic reactivity of the brain is affected by separation." Hofer is preparing to begin brain analysis of the separated rats to find exactly which brain activities are involved.

Francis V. DeFeudis at Indiana University is already involved in such work with socially isolated mice--mice kept alone in cages and allowed no physical or social contact with other animals. As with Harlow's monkeys, this procedure produces profound changes in behavior. In the isolated mice this has resulted in the classic symptoms of depression. The mice that came back from isolation reacted violently and aggressively to their renewed social contacts. "They would attack without provocation," says DeFeudis, "and their attacks--in mouse terms--were irrational. They were not fighting for food or water or space in the cage. They were fighting to kill."

Analysis of the brains of these animals showed that the psychological changes were accompanied by biochemical changes. "There is an easy and scientifically confirmed explanation," says DeFeudis. The chemicals that act in the nervous system to suppress abnormal behavior simply are not getting into the system in sufficient quantity. In other words, according to DeFeudis, social isolation significantly lowers the nervous system's capacity to produce chemical inhibitors. He found, for example, that the brains of isolated mice lost much of their capacities to metabolize glucose and manose, two forms of sugar that generate energy in the brain's biochemical reactions. ("I suspect that's true," says Hofer.)

Once brain chemistry has been altered, it is likely that animals will react differently to administered drugs. DeFeudis injected amphetamine into both isolated and socialized mice. The animals were genetically identical but they had significantly different biochemical reactions to the drug. The brains and nerve endings of the isolated mice absorbed more amphetamine. The isolated mice metabolized the drug much more slowly than did the others. Also, the amphetamine in the isolated mice further depressed the abilities of their brains to metabolize the glucose energy source. This, says DeFeudis, did not occur in the mice in colonies.

Do these reactions in mice and monkeys also happen to humans in similar situations? Says McKinney: "This is a kind of way to look at biological changes that might accompany separation in humans." DeFeudis is more explicit. "It may not be especially flattering," he says, "but the biochemistry of mice and men is very similar. We have not yet done the experiments which would confirm our findings in humans, but the apparent analogies to human behavior are overwhelming."

The results have immediate application to men in cages, says DeFeudis. "This shows how stupid it is to punish prisoners by throwing them into isolation. If their problem is aggressive and antisocial behavior," he explains, "the isolation will only aggravate the behavior. They come out of isolation with even less ability to control themselves."

DeFeudis does not restrict his speculation to solitary confinement of prisoners. He says human psychotics may be creating their own isolation and carrying it around with them. They develop an ability to shut out any stimulation from social contact. "All of us have this ability," he goes on. "We can close off much of the stimulation from the world around us when we are reading intently or concentrating on a piece of work. Psychotics seem to develop this ability perfectly--they are so isolated as the mice, even though they may be in a crowd."

The research also may have applications for work with drug addicts. Addicts, DeFeudis says, may be seeking isolated environments or creating their own isolation in much the way that psychotics create theirs. The result is that they respond very differently to drugs than do normal persons. For example, he says, troops who came back from Vietnam with drug habits usually lost the habit as soon as they were established again in their normal environments. DeFeudis likens the combat situation, where the drug problem begins, to a form of isolation. A similar situation is seen in drug rehabilitation programs. An addict goes through withdrawal and is physically cured but then picks up the habit again upon returning to the drug cluture. DeFeudis also suspects "that for many addicts, criminals and psychotics, we will find that the biochemical aspects of their problems may be the result of prolonged and increasingly successful attempts to withdraw from the environment." An inability to cope with the stimuli of their environment, he says, may have inspired them to shut out the stimuli, to create the isolated environment that produces the biochemical basis for addiction and psychosis.

Most researchers agree that changes in the environment produce changes in the biochemistry of an individual, but not all are ready to draw conclusions about humans from work with animals. Bruce Welch of the Maryland Psychiatric Research Center in Catonsville, Md., says the response of animals to all kinds of drugs changes after isolation, but we shouldn't be too extensive in extrapolation to humans.

The picture is just beginning to smooth out, he says. Isolation seems to be accompanied by a slowing down of turnover and release of what are believed to be neurotransmitters. But most of these substances, he warns, can work either as inhibitory or excitatory factors. Work with animals shows that isolation produces hyperexcitability and increased response to stimulation. Some electrophysiological work with experimentally isolated humans, on the other hand, shows a slowing of brain activity. This, says Welch, indicates a state of decreased arousal.

"Changes in the environment," Welch concludes, "do result in important changes in the brain. As we use animals to study these changes more carefully, we might develop a tool to learn how the brain works, we might learn how biochemical changes affect mental processes. But it is still too early to draw any similarities with humans."

CONSCRIPTION AT SEA
Bruce Wallace

Here are two interpretations of current "naval relations" with porpoises and other sea-going mammals.

War and preparations for war are infected by the macabre. Professional satirists recognize this and exploit it. No sooner had the Strategic Arms Limitation Treaty discussions ended than Art Buchwald, tongue in cheek, described what a boon this treaty would be to munitions makers and armament suppliers. Boon? Yes, indeed, because all of the arms not prohibited by the treaty now would have to be invented and/or perfected. Less than two weeks after the publication of Buchwald's cynical article, the then secretary of defense, Melvin Laird, appeared before a congressional committee to argue precisely that point. The arms-limitation treaty, rather than inviting a reduction in defense spending, called for an increase, because new items not covered by the treaty were now in need of research and development. Satire preceded reality by a dozen days or so.

That tragic events can hinge on logic such as this was illustrated by the late physicist Leo Szilard in his short story "The Voice of the Dolphins." In Szilard's fictional account, the Soviet Union and the United States are confronting one another over Middle Eastern oil. Russia, to show that her demands must be met, has submitted a list of twelve American cities that are to be destroyed one by one; to avoid unnecessary bloodshed, however, an advance warning is to be given so that inhabitants can be safely evacuated before the destruction of each city. In reply, the United States threatens to annihilate two Russian cities for every American one destroyed.

"To this," Szilard's story continues, "Russia replied in a second note—a

Copyright © 1973 by Saturday Review Co. First appeared in *Saturday Review*, March 1973. Used with permission.

note of unprecedented length—that if America were to demolish two cities in Russia for each city that Russia might have demolished in America, and if Russia were to demolish two cities in America for each city America might have demolished in Russia, then the destruction of even one city would trigger a chain of events which would, step by step, lead to the destruction of all American as well as all Russian cities. Since clearly America could not possibly want this result, she should not make such a threat of 'two for one' and expect it to be believed."

There is, as the Russian message implied, a predictability to warfare: events, once started, leapfrog in a seemingly ordained pattern. Consequently—and now I come to the point of my story—when the navy utilizes marine mammals as serious military allies, one can predict with considerable accuracy the course of subsequent events. Sea lions, pilot whales, porpoises, and killer whales are all being trained for various tasks by the Marine Life Sciences Laboratory, a division of the Naval Undersea Center in San Diego.

Consider the use of armed porpoises trained as guards against enemy frogmen. Upon spotting a frogman, the porpoise sends a signal to a sailor in a nearby monitoring vessel. Should no friendly frogmen be authorized in the vicinity, the sailor radios back, "Kill!" The porpoise then drives the knife that is strapped to its snout into the enemy frogman. Porpoises trained in this way were used not long ago in and around Haiphong Harbor. It is not difficult to imagine an explosive device in place of the knife. Unwittingly, the porpoise then becomes a sacrificial mine. Another project that has been reported involves whales. Wearing special harnesses, they have been trained to recover objects (torpedoes, for example) that have been lost at sea.

Judging by the few published accounts, Operation Porpoise and Operation Whale (my own terms for these classified projects) have been successful, not because porpoises and whales share the American political ideal, but because they can be trained under an appropriate system of punishments and rewards. What Americans have done, then, we must expect our enemies to do also, or so goes the premise upon which war logic is based. And if an enemy were to train porpoises for wartime duties, what should our defenses against them be? And how are we to know an enemy porpoise from one of our own or an innocent byswimmer?

We can assume that panels of learned men have already been assembled to ponder these questions. We can assume, too, that their charge as a committee was to recommend a short-run solution. They will have been asked about the feasibility of destroying porpoises and whales throughout tens of thousands of square miles of the sea, throughout a theater of operations. Just as their colleagues from the Forestry Service were consulted about the defoliation of Vietnamese jungle, and just as their colleagues from the Meteorological Service were consulted about rainmaking for military purposes, so will animal physiologists and marine ecologists be asked to plot the wholesale destruction of porpoises and whales. Is there, for example, a poison that might be spread on the surface of the sea that would be effective in killing porpoises as they surface to breathe?

One is relatively safe in predicting that panels have begun discussing the practicability of wiping out porpoises and whales, because to fail to do so would demonstrate gross incompetence on the part of our strategists. For every weapon there must be a counter-weapon; for every strategy, a counter-strategy. Hence, by the peculiar logic of war, plans must have been laid for the destruction of our new seagoing allies.

In the past, animals used in warfare shared two similarities: they were domesticated, which means that there were enough back home to replenish their fallen counterparts; and they were, by and large, clearly identifiable. It was the charging steed. Or the police dog leaping from one shell hole to the next. Or the homing pigeon carrying secret messages back to headquarters. All the enemy soldier had to do was fire away in hopes of bringing one or the other down.

Whales and porpoises, though, have few benefactors, and nobody at the moment is breeding them with any commercial intent. Also, they live in a natural environment, not on a battlefield; consequently, the enemy is unable to distinguish trained from untrained animals. Furthermore, these particular carnivorous predators occupy the top echelon in the marine community. The whole system beneath them depends on their continued existence. Their wholesale destruction could have ramifications far beyond the imagination--or concern--of most admirals. As on the land, where the diversity of plant life depends on the presence of certain herbivorous mammals and the diversity of these mammals depends in turn upon their predators, so the situation might be in the oceans. What, however, corresponds to a meadow in the marine community? Would the elimination of porpoises have consequences that would affect the composition of marine plankton? We are not sure how delicate the balance is or how easily it could be upset. However, assurances that there is no need for concern are as worthless as they are expectable.

How can a nation escape from the dilemma of using marine animals for allies and then, in fear that they may be used by others, of destroying ocean communities? One might extend the notion of bacteriological warfare, which has already been "outlawed," to include biological warfare of all kinds.

As a matter of practical importance, I would suggest prohibiting the use, for military or other aggressive purposes, of undomesticated animals, as well as animals that cannot be easily recognized as belonging to "the enemy" or to "the wild."

Such matters need prior thought; they cannot be resolved once events have been set in motion. Too many search-and-destroy missions have destroyed all human beings because a few were suspected of being legitimate targets. Too many saturation-bombing missions have been flown because pinpoint missions were difficult or impossible. Too many people have been identified as the enemy only because they were already dead. If large-scale destruction of human beings is accepted in warfare, how can the wholesale destruction of marine life be avoided except by not involving their members in mankind's problem? The steps to declare wild species off-limits should be taken now; otherwise they will not escape the macabre logic of war.

*

TUNING IN ON PORPOISES THAT "WORK" AND "TALK" WITH PEOPLE

Ringed buzzer is used by the Navy to test dolphin ability to do man's work. Acoustic buzzer in center is turned off by the dolphin when he butts ring.

What has come to be known in some circles as the U.S. Navy's "dolphin caper" got into print again last spring. The perennial story hints darkly that, as part of a "top secret" program, our wet-suited men in blue have been training dolphins and other sea-going "friends of man" to ram enemy ships with explosive charges--like flippered kamikazes.

An added fillip this time was a report that a platoon of six bottle-nose dolphins (according to unidentified "military sources") had been put together at the Naval Undersea Center (NUC) laboratory in Hawaii, and trained in the murky business of killing communist frogmen by means of switchblade-like weapons attached to the beak. The report, according to the "sources" claimed that the sea mammals had

Reprinted with permission from *Science Digest*. Copyright © The Hearst Corporation, February 1973. All rights reserved.

After pressing grabber claw against target missile (right), pilot whale Morgan disengages mouth piece which is held clenched between his jaws. This activates gas system that inflates a balloon and floats torpedo to surface. Porpoise, below, takes a line from diver rescue reel, which he will take to a lost diver, guided by sound of a tin cricket. The line is a means of guiding the diver back to home base.

been used to protect our base at Cam Ranh Bay, Vietnam, from underwater guerillas.

Those "military sources," say some Pentagon spokesmen, could be communists. For years, they point out, the Red press has accused us of using divisions of skunks to flush troops out of pillboxes, squadrons of bats to carry disease germs and incendiaries into their hideouts, rabid foxes to spread rabies and a lot of other far out stunts that one officer refers to as "opium dream nonsense."

Communist propaganda or no, every time another such report circulates, the roof falls in on the Navy; mail-bags bulge with horrified complaints from animal lovers, and the Navy issues denials, or explanations of what it is doing in its sea mammal program.

This time, Harris B. Stone, director of the Navy's Research, Development, Test and Evaluation (RDT&E) program pointed out to <u>Science Digest</u>--somewhat wearily--that none of the Navy's deep-diving work with sea mammals has ever been classified.

What about Cam Ranh Bay? Did they have dolphins there? Yes, the Navy admits.

Was it classified? Yes. "But," says Stone, "not only because of the work performed by the animals. When you're in an operational area doing experimental

Navy's pilot whale, Morgan, is rewarded with a fish snack after a successful missile retrieval. His harness "uniform" carries electronic communication gear.

work, it's automatically classified. We've taken these animals to many operational areas to check them against salinity levels, pollution levels and other environmental factors. In Southeast Asia we had a lot of things to check--can we transport porpoises over long distances? How do they adapt to new environments? Everybody talks about their surveillance and detection capability. Is it really there and, if so, will they work in an environment they don't normally inhabit? Cam Rahn Bay, where the Navy has facilities, was a good place to get new data like this."

What about blowing up ships and killing underwater guerillas?

"Let's get that straight," says Mr. Stone. "Any idea of self-destruction of these great animals is personally obnoxious to me and to anyone who has worked with them. What's more it would be inexcusably wasteful. It takes three months and costs upward of $50,000 to domesticate one animal before he's even ready to learn a simple task. We have easier, more reliable ways to detect and destroy operational targets with foolproof mechanical devices."

All this doesn't mean that the fascinating NUC sea mammals haven't been trained in operational duties at NUC, and deployed to the field. Some achievements have been highly successful.

Last September, the Navy trained two of three whales to make deep ocean recoveries of objects such as experimental torpedos and other inert test ordnance.

Morgan, a 1200-pound pilot whale, and Ahab, a 5500-pound killer--working out of NUC's laboratory at Kaneohe Bay, Hawaii--made dives to 1000 feet routinely. This is twice the depth achieved by sea lions in similar recovery duty.

Morgan, in fact, descended on one dive to a depth of 1654 feet.

The system, dubbed "Deep Ops," uses a hydrazine gas generator device attached to a mouthpiece that is grasped by the whale between his jaws as he descends to the ocean floor. An acoustic pinger attached to the target guides the whale.

A grabber claw, when pressed against the target by the whale, locks the lift device onto the torpedo. The hydrazine system is activated when the lift device detaches from the mouth-piece. The gas inflates a balloon, which floats the torpedo to the surface where it can be loaded aboard a recovery boat.

Morgan demonstrated that pilot whales can be used with the system to at least 1600 feet, and the Deep Ops project has shown potential for using other species of whales for recoveries to more than twice that depth.

During training, Morgan proved the more reliable and controllable of the whales. The first time he was released in the open ocean after training began he returned voluntarily to his pen. Not so with the killer whales. In fact, one of them went AWOL during early training and never came back.

The Navy demands a high degree of reliability on the part of its sea animals. Each trainee must remember a given task and repeat it until it is accomplished the same way each time. Orders are given by vocal as well as radio signals. The animals wear a radio set attached by harness as part of the "uniform."

The killer whale apparently is not as reliable and predictable as the pilot whale, Stone reveals. "We know what we can do with the pilots. Both types will perform a given task, but we don't know what else a killer whale may do."

This was evident, for instance, when trainers introduced their charges to the mouthpiece of the torpedo lifting device. Whales like to bite on things. As a result, the Navy developed a whale teething ring for training.

"The whales liked it so much," says Stone, "that conditioning them to carry an object in the mouth was easy. Getting the killers to release it was something else. We'd say, 'Good boy--now let's have it back.' Instead, the whale would dash off, toss it around and cavort with it. We'd get the ring back when the whale was ready.

"The trick for us is to learn what a particular animal likes to do, and adapt that to behavioral training so the animal also will do something of value for us."

Currently, something of value includes retrieval of test weapons. "This is expensive hardware," Stone explains. "To insure that it meets specifications, it is test fired as it comes off the line. But you're not going to throw it away.

"We used to retrieve such materials with divers--two to a missile. Then we needed two more divers halfway down to keep an eye on the first pair. Topside, we'd need a recompression chamber and medical personnel--an expensive operation."

Human divers can't work deeper than 200-300 feet. Dolphins go readily to 1,000 feet. Morgan worked well at 1,650 feet, and the Navy believes that larger species might be trained to dive to 6,000 feet.

While no one denies the intelligence of the large-brained cetaceans (dolphins, porpoises and whales) there's plenty of disagreement among experts on just how intelligent "intelligence" is. One major mission of the Navy's $2,000,000-a-year program is to pin this down and separate fact from fiction.

Mr. C. Scott Johnson, who heads up NUC's marine biosciences department, recently explained this to a Wall Street Journal correspondent: "The guy on the street wants to believe these animals are as smart as he is. If you tell him it isn't so, he wants to punch you in the nose."

Part of the reason for this can be laid to the published works of John C.

Lilly, who reported some startlingly anthropomorphic characteristics of dolphins during research with them through the 60s. Lilly and his co-workers treated the animals like people, lived with them in a special pool in a Virgin Islands laboratory, and claimed a comradeship and beginnings of inter-species communication that sounded more like Dr. Doolittle than science.

For two and a half months, a Lilly researcher named Margaret Howe shared the "wet room" with a cetacean named Peter Dolphin, and gave him daily lessons in English. According to Miss Howe's report, before the lessons were finished, Peter was making humanoid sounds instead of just clicks and whistles. Occasional sounds like "oie", for instance, were interpreted as delphinese for "boys". Tapes of the "conversations" purported to prove dolphin progress.

Navy scientists who have heard the "MARGIT AND PETER" tapes take a less optimistic view of Lilly's conclusion. One delphinologist claims he listened carefully, heard no human sounds, took a glass of sherry and listened again, with no better result. He admits that Lilly's work fascinated him. "I like Dr. Seuss, too," he adds.

Dolphin vocablulary aside, the Navy now knows that marine mammals can be used in a human-animal "buddy system" that is highly effective--and startling enough. This was proven during Sealab II operations, when the animals were trained to dive and bring messages and tools to divers on the ocean floor. One of the most exciting things to Stone was the ability of the porpoises to home in on a lost diver and bring him a line that would guide him back home.

"We gave the diver a metal cricket and if he got lost, he simply clicked the cricket. The animal was trained to home in on him at this signal and bring him a tether anchored at the base he left."

Many other sea mammal capabilities are still to be utilized, but the Navy has high hopes they will be one day. The animals are able to navigate and locate food without relying on eyesight; instead they project sounds and interpret the echoes. Study of this "echolocation" capability has already helped the Navy to improve its own sonar devices, which operate on a similar principle.

"One of the toughest jobs we have," says Stone, "is to locate the exact point at which an object is lost--an underwater vehicle, a disabled submarine, a diver who may be in trouble.

"I can see using marine mammals to mark these places for rescue operations. We feel that with the biological sonars inherent in the porpoise and the whale, they may be able to do the job for us. If we can train them to locate objects with precision, they might take a marker down and thus tell us: 'Here it is'."

This may happen before long. In a 1967 experiment, a blindfolded porpoise was found capable of finding, through echolocation, a 30 centimeter target. It was a round disc of 0.22 cm-thick copper plate suspended by a paddle. It was among targets of other materials, including aluminum plate. The animal was able to make the distinction.

What about actually "talking" to people? Sorry, say the Navy investigators. Recently, Dr. David K. Caldwell and Melba C. Caldwell studied the whistle-like sounds of dolphins for the Marineland Research Laboratory and the Office of Naval Research, and report that the dolphin whistle transmits only general information --its location to another dolphin or a degree of excitement or fear.

The scientists conclude that man will never be able to carry on any more lucid a conversation with a dolphin that he can with a dog.

*

THE SCARRED SPECIES
Patricia McBroom

Konrad Lorenz's theories on the innate aggressiveness of mankind (see "What Aggression is Good For") have not gone uncriticized.

Few issues bring the anthropological world to a boil more quickly than the question of man's innate aggression: Are humans born ferocious and destructive or do they learn to be that way?

The issue flared most recently with publication of two books--Konrad Lorenz's "On Aggression" and Robert Ardrey's "The Territorial Imperative"--and has been raging ever since.

Both books argue an aggressive drive in humans inherited from the evolutionary past, and their popular success caused considerable distress among cultural anthropologists who feel the argument is highly misleading.

In the cultural view, whatever man is he learns to be, since the human capacity to learn overrides any possible remnants of instinctive behavior.

In September, anthropologist Ashley Montagu replied to Lorenz and Ardrey with a collection of essays, by 14 authors, entitled "Man and Aggression." In his own chapter, Dr. Montagu writes, "The myth of early man's aggressiveness belongs in the same class as the myth of the beast...the myth of the jungle...and the myth of innate depravity or original sin.

"What we are unwilling to acknowledge as essentially of our own disordering in the man-made environment, we saddle upon 'Nature'." It is very comforting,

---

Reprinted by permission of *Science News*, © 1973 by Science Service.

says Dr. Montagu, and it successfully diverts attention from the real source of human destructiveness--namely, false and contradictory values operating in an overcrowded, highly competitive and threatening world.

Dr. Montagu's true antagonist is not Robert Ardrey, a writer, but Dr. Raymond Dart, the South African anatomist who influenced Ardrey.

It was Dr. Dart who in 1924 discovered the fossil remains of the two-million-year-old prehuman ancestor, Australopithecus. Since then he has been extrapolating from the physical evidence to create a vision of the kind of creature that bred humankind.

"These are the ogres of the fairy tales," says Dr. Dart. "They are people who would grind your bones to make their bread."

Man earned the mark-of-Cain when he turned from the herbivorous pursuits of other primates to become a carnivore, says Dr. Dart, and the "blood spattered, slaughter-gutted archives of human history" testify to the presence of a blood lust. "Man is and was a killer. He got on his hind legs so he could kill with a club."

It is worth noting that Dr. Dart refers specifically to the ferocity of pre-humans, not to an aggressive drive inherited generally from the animal kingdom. Much of the talk about evolutionary drives attempts to explain human aggression by the fighting behavior of wholly unrelated species--monkeys, rats and birds--while glossing over two million years of human evolution, during which time many species of human-related primates arose, roamed the earth and became extinct.

It is speculation about the behavior of these protohumans which underlies much of the current controversy. What in their nature, what forces of evolution produced man, the mass murderer, when even his closest living relatives, the great apes, are quite peaceful creatures who rarely kill or even fight against each other?

A good deal has been learned over the past decade about this human ancestry --Australopithecus, Homo erectus and Neanderthal man, in order of their appearance on earth.

Australopithecus includes several species of bipedal primates whose members have so far been found only in Africa. Dr. Dart found the first representative in South Africa, but it was not until Dr. Louis S. B. Leakey uncovered an Australopithecus fossil from the Olduvai Gorge in East Africa in 1959 that the creature could be dated. Radiocarbon dating revealed an age of two million years, adding another 600,000 years to human development and allowing fossil remains such as Java and Peking men to fall in their proper time scale.

At least one of these Australopithecines was a vegetarian, but he was an evolutionary deadend, helped to extinction, it is believed, by his meat-eating cousins.

As evidence of carnivorous habits, the broken skulls of baboons have been found associated with Australopithecine bones, suggesting that the creature ate the brains.

But whether Australopithecus was a true predator or a scavenger of already killed meat is not yet established. "There certainly is evidence of carnivorous and predatory behavior," says Dr. F. Clark Howell, of the University of Chicago, an authority on early man, "but the question is, how much of the total evidence suggests this way of life."

An answer, says Dr. Howell, would have bearing on the issue of man's innate aggressiveness and on the strength of that drive, if one exists. Hopefully, a group of American anthropologists now in Africa analyzing the evidence will determine how predatory the human ancestor was.

From this pool of meat-eating, upright primates came the first known species of man, <u>Homo eretus</u>, about half a million years ago. <u>Homo erecutus</u>--which includes Java man in Southeast Asia and Peking man in China--spread around the world. He was a hunter.

But again, "this was a very different kind of carnivorous behavior from other animals," says Dr. Howell. Females did not hunt, indicating that the predatory behavior was rather specialized in humans, as contrasted to such animals as cats and wolves.

In short, the mark-of-Cain thesis rests on ambiguous biological evidence. As Dr. Howell points out, man has a digestive system fit for vegetables and must cook his meat to break down the proteins. At the same time, he has an adrenalin pattern suited to the hunt.

The first <u>Homo sapiens</u>, Neanderthal man, appeared in several places about 100,000 years ago. According to the most recent fossil evidence, he evolved into modern (Cro-Magnon) man in the Middle East but not in Europe. In migrating throughout the world, modern men may have overtaken and destroyed European Neanderthals who had failed to evolve.

Neanderthal man buried his dead, built dwellings, left evidence of an artistic sense--and also indulged in cannibalism, eating the brains of fellow creatures apparently to incorporate the dead man's spirit.

The step from eating baboon brains to eating human brains seems a short one. And if the two are derived from the same bloodlust, as Dr. Dart implies, then there would be reason to believe in some extraordinary ferocity in humans, since other animals do not eat their own kind. But lacking clear evidence of cannibalism before Neanderthal times, it would appear that more than a million years of evolution had to occur before cannibalism was possible and that it stems from man's central characteristic as a species--his symbolizing brain.

Between the time of <u>Australopithecus</u> and Neanderthal man, the brain underwent rapid changes, gaining the large neocortex which was to provide the human species with its ability to manipulate symbols.

All human groups, including the most illiterate and primitive, think in symbols.

The brain is so constructed that emotional impulses--which man shares with other animals--flow constantly through this symbolic structure in the neocortex. The neocortex, in turn, sends constant impulses to emotional centers. Except in unusual cases, human emotions are not simply experienced at an unthinking level. They are channeled through symbols--words, ideas, convictions, religious, social and personal beliefs, all of them abstractions from the concrete world experienced by other animals.

"The emergence of the ability to symbol," says Dr. Leslie White, University of Michigan anthropologist, "resulted in the creation of a whole new world which contains, directs, controls and regulates basic drives." In effect it cut man loose from instinctive behavior.

The product of this capacity is culture, says Dr. White, and he adds that it is futile to try to get at biological nature "by looking through the prism of culture which refracts everything."

The essence of the cultural argument is that if humans can talk themselves into eating each other for spiritual reasons, damming up their sexuality for moral reasons, killing their brothers for political reasons and going to their own deaths for ideological reasons, where is the rationale for believing in an innate aggressive drive unmodified by those same symbolic processes?

"The very fundamental drive of sex is controlled by non-biological forces," says Dr. White, "The same is true of aggression."

Whether or not the human symbolic processes are completely arbitary, or themselves the product of evolution, concerns a group of linguists looking for universal patterns in language.

Dr. Noam Chomsky, a linguist at the Massachusetts Institute of Technology, believes symbols do have biological roots and that the kind of language and culture which humans developed over a million years was subjected to the forces of evolution.

Anthropologists take it for granted, says Dr. Chomsky, that humans can devise any kind of culture, that there are no limits. But he says, "I am morally certain that if people began understanding culture, they would discern very strict limitations."

Whether or not Dr. Chomsky is right about a biological endowment shaping culture, his theory takes even the inheritance argument back to culture. As Dr. White says in paraphrasing Alexander Pope, "The proper study of man is not mankind, but culture."

PART VIII
NATURAL HISTORY

THE BISON'S WOE
Kevin P. Shea

Who is a friend to the buffalo?

Reprinted by permission of The Committee for Environmental Information, Inc. Copyright © 1971 by The Committee for Environmental Information, Inc.

While eradication has always been a favorite word in the vocabulary of both state and federal agricultural officials, it is usually met with a chilly reception on the part of many ecologists. Past attempts to eradicate such pests as the fire ant, the gypsy moth, hog cholera, and various other insects, weeds, and diseases have met with mixed success and, almost without exception, have encountered strong opposition from some segment of the scientific community and the general public. The issues usually center around the high environmental costs of such programs as opposed to their uncertain benefits.

The latest controversy over a U.S. Department of Agriculture (USDA) eradication program is unusual in that it is being led not by a public interest group but by another govenment agency: the Department of the Interior's National Park Service. At issue is a herd of 700 free-ranging, naturally regulated, and totally wild bison.

The problem is that after many years of trying, the USDA feels it is "on the downhill side" of an eradication program designed to rid North American livestock herds of a disease known as brucellosis (undulant fever in humans). In the past the disease has been costly both in terms of losses of livestock and in terms of human health. Unfortunately, the Yellowstone bison are known carriers of brucellosis: as of 1964-65, infection rates were as high as 59 percent in some herds. Although the brucellosis organism doesn't seem to affect the bison, the USDA officials feel that this reservoir of the disease jeopardizes the success of their eradication program. At present the USDA claims that because of its concentrated efforts, 61 percent of all counties in the U.S. are brucellosis free, and by 1975 they will have achieved total eradication.

To insure against reinfection of livestock in the Yellowstone area, the USDA has offered the Park Service a plan to eliminate brucellosis from the wild bison. It involves trapping as many of the animals as possible, testing them for the disease, and killing those found to be carriers. The brucellosis-free bison would then be held in pens until all of the untrapped animals were killed on their ranges; a process that could take two years or more. At the very least, according to Park Service estimates, this would reduce the wild herds by 85 percent.

The Park Service, for a number of reasons, has rejected the USDA plan and countered with its own proposal. Since the bison rarely stray out of the Yellowstone boundaries, the Park Service has offered to routinely monitor the distribution of the bison to determine if any animals are in areas where they might move onto ranges occupied by domestic cattle. If animals are found in such areas they would be destroyed as would animals straying over the park boundaries. If domestic cattle stray into the park they would be impounded and tested for brucellosis at the owners' expense.

Both plans may be academic since there is no evidence to show that Yellowstone bison can transmit brucellosis to domestic cattle. While it is virtually certain that if bison and cattle were penned together, or if cattle were injected with the brucellosis organism from the Yellowstone bison, the cattle would become infected, no bison in the Yellowstone area have ever been shown to transmit the disease to domestic cattle adjacent to the park for more than ten years.

Another factor makes the situation even more complicated. Elk populations in the Yellowstone area are known to have a brucellosis infection rate of about 1.5 percent. The USDA is convinced that the elk are being infected from the bison and that if the bison were brucellosis-free the disease in elk would die out. In other words, they believe that there is little or no elk to elk transmission, a situation which would perpetuate the disease in elk and possibly even be a source of reinfection for the bison.

enforcement and conduct neverending disease detection programs along with costly vaccination programs to increase resistance to infection. Disease eradication in animal populations has been successfully employed on a number of occasions, and the elimination of burcellosis is equally sound.

The plan suggested for eliminating this disease from buffalo (bison) at Yellowstone National Park was selected after considering several alternatives. The plan basically calls for corralling and removing surplus (diseased) animals. Both of these procedures have been practiced for many years in the Yellowstone herd by the National Park Service. In fact, the herd size was reduced by slaughter from 1,500 to 500 in 1954-56 and from 900 to 300 in 1961-66. Both burcella-infected and healthy animals were included in these herd reduction programs. Under the plan for eliminating brucellosis, the herd would probably be reduced from approximately 700 animals to about 350 animals, but the less productive animals would be removed thereby providing for a healthier and better herd.

Similar plans to eliminate brucellosis from bison herds have been used successfully in other herds under both private and public ownership. As you know, bison are not an endangered species and are even raised commercially in some states. Your support for improving the health status of these fine animals would be welcome.

Sincerely,

Donald Miller
Director
Programs Development and Application

UNITED STATES DEPARTMENT OF THE INTERIOR
National Park Service
Washington, D.C. 20240

April 16, 1973

Honorable Clifford P. Case
United States Senate
Washington, D.C.

Dear Senator Case:

.....There are some comments in Mr. Donald Miller's letter to Mr. James H. McGregor that deserve further comment from us. First and perhaps our major argument with Mr. Miller's reasoning is that we do not consider "eradication" of brucellosis to be feasible, ecologically sane, or even possible, unless one would consider eradicating all possible, host species from the planet, which seems to us to be an unacceptable alternative. The organism can and should be controlled at levels acceptable to public health and economic concerns. Control does not depend upon eradicating the <u>Brucella</u> organism from Yellowstone bison, elk, deer, and so on.

The number of bison lost from the herds in the years stated by Mr. Miller is approximately correct. The first group of years, 1954-56, represents three winters (54/55, 55/56 and 56/57), and the second group of years represents five winters from 1961-66. It was our impression during those years that there were too many bison, based upon recognized tests used at that time to determine "overuse" of habitat. This was followed by a "reduction program." As it happened, severe winter weather resulted in a fairly high natural mortality that coincided with the intentional reduction program, with the result that more bison were lost from the herds than had originally been intended.

Since the last intentional management reduction of bison in Yellowstone, we have completed considerable research regarding both the bison and their habitat. Our conclusion is that prior intentional reductions of herd numbers were improper. We should not have carried out such programs, and had we known then what we know now, we would not have. The entire philosophy regarding the relationship of herbivores to their habitats and predators within areas managed for "naturalness" has undergone considerable change in the last few years. Additionally, large areas within Yellowstone National Park are now proposed for wilderness designation. This makes it illegal (happily) to construct the traps and corrals that would be required for such a program for two out of the three bison herd groups in the park.

Thank you for this opportunity to comment further on this very important matter.

Sincerely yours,

Robert M. Linn
Director
Office of Natural Science

*

ECONOMICS OF WILD STRAWBERRIES
Euell Gibbons

>   I wish I knew half what the flock of them know
>   Of where all the berries and other things grow,
>   Cranberries in bogs and raspberries on top
>   Of the boulder-strewn mountain, and when they will crop.
>   I met them one day and each had a flower
>   Stuck in his berries as fresh as a shower;
>   Some strange kind--they told me it hadn't a name.*

Prejudice prevents many people from enjoying delicious and wholesome wild food, but strangely I have never met anyone who prejudiced against the Wild Strawberry. <u>Fragaria</u> it was christened by some botanist with an appreciative nose, and there are two excellent species in the eastern part of our country, the <u>virginiana</u> and the <u>vesca</u>. Most people agree that the wild strawberry is vastly superior to the cultivated kind and is about the most delicious of all our wild fruits, the top prize for the wild food gatherer.

This is one of the widest ranging of all our wild fruits and, while it is abundant in some places, it is rare and found only locally throughout much of its range. It is more abundant in northeastern United States and eastern Canada than elsewhere, but I have picked wild strawberries in a number of states, from Quebec to Texas. On the Pacific Coast I have enjoyed a closely related species, the <u>F. californica</u>, which I have picked from the mountains of California to the artillery

---

Copyright © 1962 by Euell Gibbons. From the book *Stalking the Wild Asparagus*, published by David McKay Co., Inc.

*From "Blueberries" from *The Poetry of Robert Frost* edited by Edward Connery Lathem. Copyright 1930, 1939, © 1969 by Holt, Rinehart and Winston, Inc. Copyright © 1958 by Robert Frost. Copyright © 1967 by Lesley Frost Ballantine. Reprinted by permission of Holt, Rinehart and Winston, Inc.

range at Fort Lewis, near Tacoma, Washington. This species produces even smaller berries than the virginiana, but they are wonderfully sweet and fragrant. Once in Texas I found an open place in the center of a wild plum thicket, where a bushel of wild strawberries could have been picked from a few square rods of ground. I took my share, then informed a nearby sharecropper's family of this treasure, of which they had been entirely unaware.

Although everyone appreciates the flavor and aroma of wild strawberries, many people balk at the tedious job of picking them. The berries must be picked free of the calyx or "hull" and one must be careful to exclude leaves, sticks or trash, for the berries are small and easily crushed and picking over the day's haul would be too difficult. The care necessary to see that nothing but clean berries goes into the pail makes it full slowly, especially since one is under constant temptation to take a toll between plant and pail.

Early last summer I drove to an abandoned orchard on a southern slope where strawberries like to grow. The season had been perfect and the berries were so thick they covered the ground, looking like a red carpet unrolled before me. I could sit down and pick a quart of berries without moving. I stayed there all day, the strawberries sufficing for my lunch. A sudden shower caught me far from the shelter of my car, so I just kept picking.

> You ought to have seen how it looked in the rain,
> The fruit mixed with water in layers of leaves,
> Like two kinds of jewels, a vision for thieves.*

The returning sun soon dried my clothes and the berries seemed brighter and fresher than before. The day was a revel in beauty, flavor and aroma, and at its close I felt that I had spent few more worth-while days in my life.

My neighbor raises strawberries in his garden. When he found how I had spent my day, he felt sorry for me. He thought it a shame that I had driven twenty miles and spent the day obtaining only twelve quarts of berries, while he had been able to gather the same quantity of cultivated berries in an hour from his own back yard.

A friend who shared our shortcake that evening, although he appreciated the superior flavor of the wild berries, felt bad about me having spent my day at such tedious labor while he was enjoying a game of golf. He wondered aloud if it really paid to pick wild berries when one considered the labor involved, the amount of fruit obtained and the price of berries on the market.

I could have argued with these two men, but I didn't. It would have been easy to have reminded my neighbor of the many hours of spading, planting, weeding, mulching and runner-pinching he had devoted to his little strawberry patch, while I had done nothing for mine except pick them. He had put out hard cash for plants, tools, fertilizer, mulching materials and taxes on his land, while I had invested nothing but a pleasant drive in the country. To my friend, who was worried about the economics of berrypicking, I could have pointed out that my day had been far less expensive, no more strenuous and considerably more profitable than had his own day at the golf course.

I could have argued the economics of wild strawberries, but it would have been pointless. The truth is that none of us spent that day seeking economic gain. All three of us had been searching for something which is hard to put in words. In the poem at the beginning of this chapter, Robert Frost has beautifully symbolized the elusive treasure we were hunting as a strange flower without a name. May-

* Ibid.

be my neighbor saw this flower growing in the corner of his garden, and my friend might have glimpsed its color in the rough beside the fairway. I found it mingling its fragrance with that of the wild strawberry in an abandoned orchard. I felt no need of economic profit to justify my having spent a day in its neighborhood.

> Who cares what they say? It's a nice way to live
> Just taking what Nature is willing to give
> Not forcing her hand with a harrow and plow.*

It hardly seems necessary to tell anyone how to use wild strawberries, but you do have a choice of many excellent ways in which to use this best of wild fruits. When picked far from roads and human habitation, they are safe and delicious to eat straight from the plant. Carried into the house or camp and eaten with whipped cream or even rich milk, they are superb.

There is no need to slice wild strawberries when preparing sugared berries. Just put the washed berries in a jar or deep bowl and sprinkle on 1 cup of sugar per quart of berries. This causes the juice to flow and the berries are soon swimming in a sirup made of their own juices and sugar. Cover the jar or bowl and set in the refrigerator until used. I have kept sugared berries in the refrigerator for more than two weeks and even then it was gluttony, not spoilage, which caused them to disappear.

These sugared berries are a joy to eat with cream or milk and, as a topping for ice cream, they are a pure delight. A shortcake made with sugared wild strawberries will cause you to lose your taste for the ordinary kind.

For an easily made family-sized Strawberry Shortcake, take 2 cups of commercial biscuit mix, add 2 tablespoons of sugar and 3/4 cup of light cream. Mix well, then turn the dough out on a floured surface and knead gently eight or ten times, or until smooth. Divide the dough into two equal parts and roll one half until it fits a 9-inch pie pan. Brush the top of it gently with melted butter, then roll the other half of the dough to the same size and place it on top of the first. Bake 15 to 20 minutes in a hot oven. The top layer will easily lift off. Spoon sugared wild strawberries and whipped cream between the layers and over the top. Generosity calls for at least 1 quart of berries and 1/2 pint of whipping cream. Cut in wedges and serve it warm, then stop envying the gods their ambrosia.

Another excellent way to use wild strawberries, either in camp or at home, is to make Wild Strawberry Roll-ups. To 1 cup of commercial pancake mix add 1 egg, 1 cup of milk and 1/2 cup of rich cream. This will make about a dozen 5-inch pancakes. Heat a lightly greased griddle to 400°, or until a drop of water will dance over its surface a few seconds before evaporating. Beat the batter until smooth, then drop on the griddle, a cooking spoon full at a time. Turn them only once, when the bottom is golden brown and the top is bubbly. As soon as they are done, fill each pancake with a generous spoonful of sugared wild strawberries and roll it up. Arrange them three to a plate, sift powdered sugar lightly over them, add a dollop of whipped cream topped with a bright berry or two and serve them while still hot.

Still another delightful dessert combines wild strawberries with cream cheese in a pie which will make you the envy of all the cooks in the neighborhood. This requires a 9-inch baked and cooled pie shell. Make this with a commercial pie crust mix or make your own favorite pastry. Let an 8-ounce package of cream cheese stay in a warm place until it softens, then blend it with 2 tablespoons of lemon juice, a little grated lemon peel and 1/4 cup of sugar. Spread this evenly over

---

* Ibid.

This is contrary to observations made in the Jackson Hole area, where elk populations have been known to have infection rates as high as 25 percent and most of the transmission is from elk to elk. Deer in various parts of the country are also carriers of brucellosis--a situation which makes eradication of the disease even more difficult.

Aside from the problem of wild carriers of brucellosis, there are some big questions to be answered concerning the nature of the disease in domestic animals. As counties are declared brucellosis-free, farmers are forbidden to immunize their stock against brucellosis, a practice that over the years has been about 65 percent effective in protecting cattle herds. The reason for this is that when cattle are tested after they are immunized, veterinarians are unable to tell whether a positive test is the result of an early immunization or a naturally acquired infection. As a result of this restriction, cattlemen in brucellosis-free counties are now building herds that are not innoculated and are thus highly susceptible to the disease. Should there be an outbreak of the disease in non-immunized herds the losses could very well be great.

Making the situation more precarious is the possibility that the disease organism, Brucella abortus, may be able to survive in swine. If subsequent studies prove this a real possibility, swine herds would represent a far more serious threat to cattle than would bison in Yellowstone Park. Adding to this problem is the fact that the disease is very difficult to detect in individual swine. Because of imperfect testing methods, individual swine may react positively in one test and negatively in a subsequent test. The best veterinarians can do at this time is declare a swine herd infected or not infected, without being able definitely to identify infected individuals.

For the time being the two agencies seem to be at a standoff. The National Park Service, while conceding the benefits of a nationwide effort to reduce brucellosis, under no circumstances seems willing to go along with the USDA proposal to reduce the bison herd and is saying so loudly and publicly. On the other hand, the USDA recognizes that it can act only in an advisory capacity and has no authority to put its plan into action however much it would like to.

The following correspondence, initiated by James McGregor of Princeton, New Jersey, outlines the arguments of the two agencies:

UNITED STATES DEPARTMENT OF AGRICULTURE
Animal and Plant Health Inspection Service
Veterinary Services
Federal Center Building
Hyattsville, Maryland 20782

March 2, 1973

Mr. James H. McGregor
100 Linden Lane
Princeton, New Jersey 08540

Dear Mr. McGregor:

Thank you for your letter regarding your views on the eradication of brucellosis from livestock. This program has made significant progress and is in the final stages of completion. The concept of eradication versus control is favored, because in the long run it is much less expensive. In order to control a disease at a low level, it is necessary to apply continuing restrictions with appropriate the bottom of the pie shell, then fill the shell to the top with well-drained sugared wild strawberries. Cover the berries with a glaze made by cooking 1 cup of the strained juice from the sugared berries with 2 teaspoons of cornstarch. Use low heat and stir constantly until it is thick and clear. Let this glaze cool to lukewarm, then pour it over the berries in the shell. Chill before serving and pass around a bowl of whipped cream for those who want it.

For some warm evening, make a Wild Strawberry Chiffon Pie, light as a summer's cloud. For a Graham Cracker Crust, crush 18 graham crackers with a rolling pin, then combine with 1/4 cup of sugar, 1/2 cup of melted butter or margarine and 1/2 teaspoon unflavored gelatin. Mix well, then press evenly into a 9-inch pie plate. Bake in a moderate oven only about eight minutes, then set it aside to cool.

Soften one envelope of unflavored gelatin in a little cold water, then dissolve in 1/2 cup of boiling water. Let this cool slightly, then mix in with 1 1/2 cups of crushed sugared wild strawberries. Add 1 tablespoon of lemon juice and a pinch of salt. Mix well and place in the refrigerator to chill. Whip 1/2 cup of heavy cream until it peaks. When the strawberry mixture has chilled till it mounds when spooned, fold in the whipped cream. Beat 2 egg whites to soft peaks. Gradually add 1/4 cup of sugar, beating till stiff peaks form, then fold into the strawberry-cream mixture. Pour into the Graham Cracker Crust and chill until firm. Just before serving, cover with more whipped cream and top with whole wild strawberries.

Cooked Wild Strawberry Jam and Preserves are so superior to those made with the cultivated variety that one shouldn't mention them in the same breath, but you can do even better than that. You may have noticed that in none of the above recipes were the wild strawberries cooked. This didn't happen accidentally; I planned it that way. The fragrance of this _Fragaria_ is one of its chief charms and this summery aroma is largely dissipated by cooking. It is possible to make uncooked jam and preserves which retain most of the aroma and flavor of fresh-picked fruit.

To make this superior kind of jam, you will need 2 cups of crushed wild strawberries made of fresh-picked fruit. Into a small pan measure 3/4 cup of water and add 1/2 cup of commercial liquid pectin or 1 package of powdered pectin. Bring this liquid to a boil and boil hard for 1 minute, stirring constantly. Cool slightly, then mix with the crushed berries. Stir it well, then add 4 cups of sugar and stir briskly for 3 to 5 minutes, or until your arm gives out. Ladle into half-pint jars and store in your refrigerator if it is to be used within 3 weeks. If you want to keep it for next winter, store it in the freezer.

Uncooked preserves are really cured in the sun. Sprinkle 4 cups of sugar over 1 quart of whole wild strawberries, and let it stand overnight. Next day, drain the berries and put the juice and any unmelted sugar in a saucepan. Just bring it to a full boil, then remove it from the fire and let it cool to room temperature. Mix the sirup with the berries in a glass dish and cover with one of the commercial plastic wrapping materials you can obtain from the supermarket. Be sure the plastic is stuck tightly to the edges of the dish all around. Place the dish in the sun. Remove the plastic cover once a day and stir the preserves thoroughly but gently so as not to break the berries.

Keep it in the sun for 4 days, then ladle it into sterilized half-pint jars and seal with sterilized lids. If it has been cleanly handled, it will usually keep well just stored in a jam cupboard, but, if you want to be certain, store it in the freezer.

You can also freeze the fresh berries, if you can find the time to pick a sufficient quantity so some can be spared for this purpose. And let me say in passing that you are very unlikely afterward ever to regret the time spent in picking wild strawberries.

You can use the same kind of half-pint jars as those recommended for jams and jellies as containers for your frozen wild strawberries. Make a sirup by boiling together for a few minutes 2 parts of water by measure to 1 part of sugar. Let it cool before using it. Use only fully ripe but still firm berries, as freshly picked as possible. Wash the berries gently in cold water, pack in the jars without crushing, cover with the sugar sirup, seal tightly with two-piece dome lids and freeze as quickly as possible. Then, any time of the year, instead of just telling your friends about the superior qualities of wild strawberries, you can demonstrate them.

All these wild strawberry products are easy to make and simple in composition. But when one of these delicacies is placed before you and you see its bright color, inhale its fragrance, taste its flavor and feel its texture, you will find, if you have picked your own wild strawberries, that it is compounded of many ingredients not mentioned in the recipes. It will bring back the warm sun on your back as you bent over the plants bearing the jewel-like berries. In it you find the grateful shade of the hickory tree under which you rested, and the old pheasant hen, who, all unsuspecting, led her brood so near. There too will be the profusion of buttercups and daisies which dotted the open field. In just a small dish of frozen wild strawberries, you can recapture, in midwinter, a long and perfect day of June weather. Some of the most precious moments of my life have been spent picking wild strawberries.

*

DANDELION GREENS I

6 to 8 quarts tender young
  dandelion leaves, well washed
  (a dishpan full)
Cold water
4 slices thick lean bacon
1 egg, lightly beaten
1/4 cup brown sugar
1/4 cup cider vinegar
2 hard-cooked eggs, chopped
Dash of salt

1. Put the dandelion greens in a large kettle and cover with water. Bring to a boil, turn off the heat and let stand 10 minutes. Drain well.
2. In a skillet, fry the bacon until crisp. Remove bacon, crumble and reserve bacon bits.
3. Combine the egg, sugar, vinegar and one-quarter cup water and pour into the bacon drippings in the skillet. Heat slowly, while stirring, until slightly thickened.
4. Remove from the heat, add greens and stir to coat evenly.
5. Serve garnished with reserved bacon bits and chopped eggs.

Yield: Six servings.

*

PADDLEFISH CULTIVATION POSSIBLE
Col. George S. Bumpas

Will the paddlefish replace fried chicken as the gourmet food of the drive-in restaurant?

Out of our prehistoric past, a fish called the "paddlefish" is now getting in the limelight. Virtually unchanged in many millions of years, the paddlefish may very well be one of the most-utilized fish of the future.

This odd creature, with ancestors predating the dinosaurs by millions of years, has been forced to adapt itself to new and changing environments or to face extinction many times in geologic history. For instance, what happened in the rivers of North America just a few thousand years ago when the last ice glacier melted? We know the paddlefish survived. Any creature so adept at survival would seem to merit our attention. Survival is something mankind needs to learn more about.

The paddlefish is one of our larger fresh-water fishes. It is the one and only species of its kind left in the world today, although it has a close relative which lives in the Yangtze River system in China. At the turn of the century, the paddlefish was found in abundance in the Mississippi River, and supported a commercial fishery which took thousands of tons of these fish annually. The boneless flesh commanded a good price; the roe was considered an excellent source of caviar, and the oils from these fish were considered very valuable. A certain type of gelatin was made from the scraps of these fish.

This unusual creature is scaleless and similar to the blue catfish in color, with a long snout, small beadlike eyes, and an enormous mouth with flapping gill

---

Reprinted from *The American Fish Farmer*, May 1973.

covers. It is now found as far south as Louisiana and Texas, as far north as Montana, and in the New York region. At one time it was reported in the Great Lakes, but it is now believed to have disappeared from those waters.

Except in some centralized localities such as the Osage River basin of Missouri, Fort Randall Reservoir in South Dakota and Lake Cumberland, Kentucky, paddlefish today have become scarce. Flood control and navigation dams on the large rivers have greatly reduced their numbers, by preventing them from reaching their needed spawning grounds. In former days, paddlefish weighing over 200 pounds were reported. Today a paddlefish is considered large if it reaches a length of 58 inches or a weight of over 60 pounds.

The paddlefish may very well be one of the most successful large plankton feeders, possibly surpassed only by the whale shark, which is one of the non-schooling ocean fishes.

Paddlefish obtain their food by swimming through the water with open mouths full of comb-like gill rakers straining plankton out of the water. If the fish are attracted by a swarm of daphnia on the bottom, they often swim over it, snouts weaving back and forth in a figure-eight spiraling motion. This type movement brings the daphnia up into open water, and the paddlefish begin to circle and feed actively through it.

When plankton is abundant, paddlefish grow very rapidly and attain a large size in a very short time--over twenty inches by the second year. They are believed to be very long-lived and are capable of attaining giant size.

The ability of the paddlefish to utilize a wide range of phytoplankton as a substantial part of their diet represents a most important link between the basic productivity of a body of water and man. Such an adaptation gives these fish access to many times the food supply of the more carnivorous groups of fishes. With increasing numbers of people in need of food and other renewable resources from the waters of this earth, we cannot afford to use the food substance necessary to raise carnivorous fishes for meat. We must find more ways to get nearer to the base of what ecologists call "food chains" or "pyramids." As an example, when conditions are favorable, dissolved minerals in the water and sunlight would produce an algal bloom. The phytoplankton would then feed such plankton feeders as the paddlefish, which in turn could help feed man, and may very well give mankind a biological method of preventing the overfertilization of a body of water from nutrient-rich sources.

Like the white amur, or grass carp, the paddlefish must have large tributary rivers, with enough water over the gravel bars, in order to spawn; they will not reproduce in impoundments which are not fed by large rivers.

Thus again, like the white amur, it should be possible to control the numbers of paddlefish in ponds and lakes; and these fish could very well help other fishes, such as the white amur, by cleaning up the plankton produced from their fecal material.

A method of artificially propagating paddlefish was first successfully developed in 1960 by the Missouri Conservation Commission, and the Commission has done research on the fish since then. Paddlefish by the thousands are hatched every year at Painted Pony Hatchery, near Marshall, Mo., which is operated by the Commission.

The first efforts to raise paddlefish on a commercial basis are being undertaken by several private hatcheries in cooperation with Windsor Marketing Corp., of which I am president. Some paddlefish are now available for sale.

There are no bones in the cartilaginous body of the paddlefish, and the flesh is exceptionally good eating. It can be smoked, fried or baked, and is known around some of the commercial fish markets as boneless catfish, spoonbill, spoonbill catfish and shovelnose cat.

The equipment anglers use in catching paddlefish is a heavy snagging rod, with a 60 to 120-pound test line. Preferred is a Penn 49 reel. Six-to-eight-ounce lead is attached to the end of the line, with one or two large treble hooks attached approximately one to two feet above the lead.

The rig is thrown across the current in the rain-swollen river, then retrieved with a slow-jerking motion, or the rig can be trolled behind the boat. When a paddlefish is hooked, it is not uncommon to hear the fishermen scream, "Fire in the hole!"

Incidentally, years ago various grab hook assemblies were baiting with dough bait and used by fishermen on the Osage River to successfully hook giant buffalo fish.

Although probably illegal in public waters of most states, there is no reason why these methods could not be used to harvest such fish as above mentioned from pay fish lakes.

In January, 1972, the American Fish Farmer & World Aquaculture News had a very good story by Sue D. Lewis on how salt was used to better catfish growth and food conservation. The predecessors of many of our present-day fresh-water fishes were salt-water species. Some became land-locked and adapted themselves to new conditions. Fishery managers were quick to take advantage of this, and introduced these fish to other inland waters. But has anyone investigated the possibility of introducing certain schooling fresh-water species, such as the paddlefish, back into the ocean? Such an endeavor could be very helpful in providing the sea with a large, boneless, and especially good-eating fish, with many valuable by-products --which could very well be an improvement over the bony, herring-like fishes which now comprise the bulk of the world's fisheries.

Wouldn't it be possible--if millions of eggs or fingerlings were gradually exposed to different concentrations of salt water--that a few of these would survive and adapt themselves, thereby furnishing mankind with a new source of food from the ocean?

When the paddlefish go up the Osage River of Missouri to spawn in the spring, they seem to by-pass other rivers such as the Niangua, Pomme de Terre, Grand, and Sac, even when these tributaries have been level full with water. This may very well be because the waters coming out of the state of Kansas contain more salt and other minerals such as magnesium. A little ocean salt could, perhaps, help prevent fungi from developing when paddlefish eggs are being hatched. The taste of these minerals may also induce fish to start upstream in the spring.

Many forms of life undergo great change in their environment without harm. For example, the Mediterranean Sea has many places where the water is over four per cent salt, yet some fishes in the Mediterranean are of the same species that are found in the Baltic Sea, where the dissolved salts have a variation of almost fresh water when the winter snow melts along the coast, and the evaporation in the summer brings the salt content back up to about two per cent in the fall.

It staggers the imagination when we truly realize what great steps could be taken in aquaculture to help millions of people at home and abroad, if experts like those mentioned above would take a long look at our living organisms like the paddlefish that have been inhabiting the world successfully for a few hundred million years.

Is it sensible, for one moment, to think that we have even begun to understand man's need for this odd animal? Who can say what obscure, odd-looking creature may some day be precious to mankind?

Antibiotics are produced from living organisms. The search for newer and better antibiotics, just like aquaculture, is in its infancy.

If we only knew that one wierd-looking animal would help us explore the stars or feed the increasing numbers of people, we would search for that little fellow. Our view of what needs to be done may not be agreed upon by others; but even if some of us make a mistake, we should be one step farther down the road to success as a whole--and perhaps future generations will have a better understanding as to what role the paddlefish will play in nature.

<u>Colonel George S. Bumpas</u> is the president of Windsor Marketing Corp., Springfield, Mo. The corporation represents several companies as distributors of outdoor and indoor recreational equipment and science and technological projects, such as Environmental Pollution Control Systems, Inc., Computerized Real Indoor Golf Corp., Chuck-Wagon Manufacturing Company, etc.

Colonel Bumpas, a Kentucky colonel, is a native of Henry County, Missouri. He was born and reared on a farm near the Osage River basin, which for many years has been the stronghold of a very large paddlefish population. Colonel Bumpas has spent his lifetime in business management and sales of various products and services benefitting humanity. He is a member of the Masonic Lodge, is a 32nd Degree Scottish Rite Mason, a York Rite Mason and a Shriner. He has devoted many years of his life toward financial assistance for the Shriners Hospitals for Crippled Children throughout North America, and for many years has been interested in the problems of ecology.

Inquiries about paddlefish should be addressed to Windsor Marketing Corp., 208 McDaniel Blvd., Springfield, Mo. 65805

†